I LIVED TO FLY ANOTHER DAY

Confessions from an extraordinary, varied flying career

George Wrigley

Amazon KDP

CONTENTS

Title Page
1 Air Experience – My First Flight ... 7
2 Mike Hill – A Man Of Character ... 18
3 Hawker Hunter – First Solo ... 32
4 Joining The Fleet Air Arm ... 42
5 Aircraft Carrier Operations ... 58
6 A Spectacular Twinkle Roll ... 89
7 Leopard Air In Malawi ... 101
8 Joys and Trials Of A Flying Instructor ... 109
9 Night Sortie at Wankie ... 117
10 Navigation Flying – Tribulation to Triumph ... 120
11 Start of Helicopter Flying In 'The Sticks' ... 129
12 Two Weeks in Chicoa–Mozambique ... 134
13 Lesley Sullivan Flies to Mukkers ... 145
14 Near Catastrophe at Mazoe ... 149
15 An American Called Jed ... 155
16 Mushumbi Pools ... 166
17 Fire Force Arrives at Buffalo Range ... 169
18 Mountain Flying Training ... 186
19 Bob Mackie Flies Me to Mukumbura ... 195
20 Hang Glider's Narrow Escape Over Durban Bluff ... 203

21 Crop Spraying in Rhodesia	212
22 Operation near Monte Casino – Mozambique	225
23 Robben Island – Close Shave	231
24 I Thought the Fishes Sleep on The Bottom at Night.	239
25 Bernard – Larger Than Life Pilot	245
26 Russian Aphid – Miracle	252
27 Crop Spraying In The Gamtoos Valley	269
Flying – Better Than Actual Work	276
29 Mercenary Pilot for the CIA	290
30 Bop Air – Start to Airline Flying	324
31 Captain, One Thing I Can't Do Is Fly	334
32 Yes, But You Don't Seem to Do Very Much	339
33 Night Flight to Thabanchu Airport	343
34 Gulfstream Flight from Abidjan	347
35 African Presidents' Wives on Corporate Flights	361
36 Poodle Popsicle Switch	365
37 DC-9 Ferry Via Iceland to The USA	370
38 Dave Kuhn – A Memorable Co-Pilot	381
39 Girl Pilot – Maxie	387
40 Windshield Anti-Ice	391
42 The Last Sun Air Flight	396
41 Intolerance to Alcohol	400
43 Turbulence and Thunderstorms	407
44 Boeing 727 Flying and Night Freight	413

DEDICATION

I dedicate this book to the pilots, engineers, friends, and compatriots who paid the ultimate sacrifice when the odds stacked against them.

GEORGE WRIGLEY

HIGH FLIGHT

Oh, I have slipped the surly bonds of earth

And danced the skies on laughter-silvered wings

Sunward I've climbed, and joined the tumbling mirth

Of sun-split clouds ... and done a hundred things

You have not dreamed of ... wheeled and soared and swung

High in the sunlight silence. Hov'ring there,

I've chased the shouting wind along, and flung

My eager craft through footless halls of air.

Up, up the long, delirious, burning blue

I've topped the windswept heights with easy grace

Where never lark, or even eagle flew.

And, whilst with silent, lifting mind I've trod

The high untrespassed sanctity of space,

Put out my hand, and touched the face of God.

John Gillespie Magee, Jr.

INTRODUCTION

This book is not a biography, or a novel, but a set of free-standing stories. Although I have incorporated them in chronological order, I have written them so that they stand on their own and readers can enjoy them as complete. However, this does lead to some repetition because readers will not understand what specific terms, like, 'a monitored approach, 'mean.

I have met several exceptional and unusual characters, undertaken many different jobs, and flown a variety of fixed-wing and helicopter types. It was my extraordinarily good fortune to have survived forty years of flying although I had many near misses during this time. A few examples are being shot at while working for the CIA as a mercenary in the Caribbean; ditching in the ocean; experiencing a tail rotor failure in a helicopter; narrowly missing flying into a hill at night and escaping under and over electrical cables. I have recounted these and many more episodes in this book, some of which were due to my, or other pilot's error, and I hope they will be of interest to those who enjoy aircraft-related stories. The reasons for many of my shortcomings have come to light, but I am not sure why I have l survived despite my foolhardy escapades.

My flying career included flying Sea Vixens off HMS Victorious for the Fleet Air Arm and instructing on fixed-wing aircraft in the Rhodesian Air Force. I also taught on helicopters in Rhodesia and was deployed on Alouettes in the bush war. I then flew helicopters on ship service from Cape Town and did crop spraying in the Bell 47 helicopter. I flew for the CIA as a mercenary, during the Contra war in Nicaragua, and finally flew as an airline captain for a South African airline. Many stories also highlight the battles pilots endure against adverse weather conditions.

PREFACE

Some years ago, Jonathan Leach came to stay with me in the Inyanga district of Zimbabwe. He came with his wife Jane to spend time in that beautiful mountainous area to write a book. We would join for lunch under a shady Canadian Maple on our lawn, and as we overlooked the spectacular panoramic scene, we would share stories from our past. After one lunch, I decided to write a flying story for Jonathan, and when he was so enthusiastic about it, I was encouraged to continue with more. Jonathan never finished his book, but rather enjoyed going for long walks in the mountains. But I am thankful for his encouragement gave me the impetus to continue.

Many people enjoyed the stories without bringing the few flaws to my attention. But when I sent a copy to Charles (Vic) Wightman, an honest friend, he mentioned many inaccuracies, and I am grateful for this. I first published this book under "Against All Odds", but with computer problems, certain edited content became scrambled. Since then, I have spent many months revamping the first edition, removing grammatical and typographical errors and correcting a few factual mistakes. I am not very good at proofreading so found it helped to listen to the book being 'Read Aloud' on Word. After that, Tish went through it all again and I must thank her for the superb job she did.

I have also changed the title because the previous one did not accurately depict the content, and other authors have used it before. I have withdrawn 90% of the photographs, the timeline, and a lengthy glossary in this updated version. You can now find most technical words on the Internet.

Once again, I am exceedingly grateful to our son Ryan, who now lives in Sydney, and who designed a new book cover for me.

ONE REVIEW OF THE ORIGINAL BOOK

I have received many emails from people who bought the first edition, Against All Odds. I have taken the liberty to include one of the reviews from this book:

5.0 out of 5 stars An aviation classic to be & a story of a life less ordinary to stir the soul.

Reviewed in the United Kingdom on 21 January 2015
Verified Purchase

George Wrigley has written an account that deserves to be ranked among the great classics of aviation history. George's life story reads like a Wilbur Smith novel, yet his understated writing style perfectly conveys the matter of fact approach of the Rhodesian "Blue Jobs". In a company where courage is taken for granted, hyperbole is scorned, and humour becomes the vehicle of choice. He writes smoothly and with the crystal clarity of the desperate deeds and passion which made the tiny Rhodesian Air force the equal of any in the world, with efficiency ratings of the scale used by wealthier countries. His accounts of places and people are accurate and penetrating, and will be of interest to casual readers, as well as military historians and people who love to hear about flying from one of the eagles of the profession. A great story from a modest man of rare talent.

1 AIR EXPERIENCE – MY FIRST FLIGHT

Percival Provost T52

It was 1960. The end of my final year at school was fast approaching, and I still had no idea what career to pursue. Having enjoyed biology and chemistry, I vaguely thought working in a fuel company laboratory would be enjoyable. Then I received a bursary offer at the local university to study for a career as a science or history teacher. However, I was fed up with examinations by that time, and thought I would instead follow a profession that did not necessitate further exams. In time, I was to learn that I could not have been more mistaken. Over the entire forty years of my flying career, I was never able to stop studying for tests and examinations.

I even thought a career in the army, would be exciting, so applied for officer cadet training at Sandhurst in the UK. But despite sailing through the IQ tests, the recruiting team did not accept me, as other applicants outshone me with their more army-style leadership qualities. Later in life, I dealt with army personnel and realised that I would have been like a square peg in a round hole in the army.

Before my final exams, I was still undecided what

career to follow when a recruitment advert appeared in the Herald newspaper inviting applicants to join the Royal Rhodesian Air Force (RRAF) and undergo pilot training. I had never considered this career as my only previous interest in aviation had been a few poorly constructed, line-controlled model aeroplanes. Many of my later friends had cycled to the nearest airport to watch aircraft taking off and landing. But sadly, some of these same people failed to meet the rigorous demands of 'stick and rudder'. This advert, which I saw by pure chance, established a career path with all its twists and turns in the years to come.

The first time I heard Air Force jets was at the New Sarum Air Force base, located on the other side of the terminal, at Salisbury Airport. The buzz of activity from the far side of the security fence, the sound of aircraft engines starting up, and the shrill scream of two Vampire FB9 fighter jets, swooping overhead for a formation break into the circuit, were an exciting introduction to an unknown world.

I saw these fighter jets the day I reported for the aptitude tests, medical examination, and interviews at the New Sarum Officers' Mess. I was among over a hundred hopefuls when I arrived there, all in jackets and ties. A few like me were still in school blazers and long grey school trousers. Once again, I sailed through the IQ tests, which were the same as I had previously tackled during the appraisal for Sandhurst.

Two young men stood out as we clustered in the snooker room, waiting for our interviews. Mike Hill, whom I remembered as one of Jameson House's borders at Prince Edward School, was one, and the other was Tony Smit, who was busy impressing anyone within hearing distance with vivid descriptions of his flying experiences. He had come up from South Africa for the interviews. In a heavy Afrikaans accent, he spoke with authority for the benefit of all and sundry, who could gain from his apparent skill and experience as a pilot. He rattled off different aircraft types, fuel consumptions, and cruising speeds. He also covered a wide range of topics from his knowledge of aviation matters. These included pitfalls to avoid if selected, and progressed to more advanced flying manoeuvres like aerobatics, spinning, side slipping, and stalling.

In contrast, Mike was casual, quietly confident, totally relaxed, and absorbed in a snooker game. He had gravitated towards a group of young men of a different calibre, who were more at ease in a pub environment, as they chalked their cues around the table. Observing him as he played, I noticed a blonde curl threatening to fall over his forehead. With a cigarette dangling from his lips, it seemed that the least of his worries was the forthcoming interview.

By the time they called me into the Ladies' Room, set aside for ladies' functions, I was in a severe state of nerves and hoped this was not too obvious. With a smile more like a grimace, I sat down in front of a long table in front of three uniformed officers, who wore wings on their chests and gazed at me with steely eyes, making their initial private assessments of me.

My stepfather had primed me for the first question, "Why do you want to become an Air Force pilot?" I knew that the wrong answer was, "Because I like the uniform." So, with as much sincerity as I could muster, I lied, "Sir, it has been a burning ambition all my life. I have always thought a military career would suit me." Other answers seemed to satisfy the panel, who said they would get back to me. But just as I let out a sigh of relief with my hand on the doorknob, Flight Lieutenant (F/Lt) Dams enquired if it was usual to wear bright yellow socks with a school uniform. To this I replied with a stutter that this practice was normal when we wore longs. But although this information was correct, I doubt whether Prince Edward authorities would have sanctioned it. In hindsight, I was taken off guard and still squirm when I think of how incongruous it must have looked.

Within a week, I received confirmation that my application was successful and was sent join-up instructions for the following month. That meant I would have to miss the Cambridge Higher School examinations. Although this suited me, the headmaster, Mr Gaylard, in his wisdom, appealed to the Air Force and persuaded them to postpone the date of my induction until I had completed these exams. I would like to have said that I put my mind to studying for these, but sadly, there were too many other distractions, and I lacked the maturity to appreciate the possibility that I might have flunked the pilot training course. I would then have needed the Cambridge qualification for an alternative occupation. Regretfully, I failed to live up to the expectation that I would repeat my previous glowing Cambridge School Certificate results.

When the day came for my induction to the Air Force, my stepfather dropped me off at New Sarum's guardroom gate. An Air Force `admin` officer then drove me in a staff car to the Thornhill Air Force base, just outside Gwelo. When I arrived there, I joined the other nine cadets of No. 14 PTC (Pilot Training Course), who had become accustomed to the harshly imposed military discipline. For me, it was a rude awakening to come straight out of school into such an unusual and unexpected environment. It was a shock to my system to be thrown directly into the deep end.

For a start, our billets were in two dormitories that were part of an abandoned building, formerly the Sergeants' Mess, built during World War II. The builders had cladded the outside with corrugated iron sheets, and the inside walls with ceiling boards, covering a timber frame. It was Spartan, with a cold cement floor. The cadets slept on a hard coir mattress on top of a steel frame bed, with a wooden laundry box at the foot. Against the walls were their bedside lockers and wardrobes. Before early morning inspection, each cadet would tuck their grey under-blankets around the mattress in such a way that the edges were razor-sharp. The sheets, pillow, and top blanket were folded flat to precise measurements and placed on the bedside locker. Each cadet was also responsible for dusting every stick of furniture and polishing his section of the cement floor to a glossy shine with red Cobra polish.

Anticipating the morning inspection, the duty cadet would stand to attention at the outside door. Here, he would wait for the inspecting officer with Colour Sergeant Paddy Malloy, an ex-soldier from the Irish Guards, now our drill and PT instructor. As Colour Sergeant Malloy prowled menacingly around, his hobnailed boots crunched across the mirrored cement floors. The duty cadet would call the other cadets to attention as they waited in dreadful foreboding. Their kit would be starched and ironed, each shoe spit and polished, belts, cap bands, and epaulettes whitened with Blanco, and brass bits glossed with Brasso. Each cadet stood ramrod straight, chest out and eyes fixed straight ahead. Stony faced, and with bated breath, each cadet prayed they had not overlooked a blemish or dust which would invoke further harsh discipline. The tiniest speck of dust incurred disproportionate punishment, which often meant having to re-polish the retired Spitfire on its pedestal in front of the admin block. Other favourite disciplines were to confine us to barracks or have us report for kit inspection at the front guardroom gate every hour on the hour.

Drill periods included further kit inspection by C Sgt Malloy. With his heavy Irish accent, he would then drill the squad at a frenetic pace with minimal time between movements or direction changes. Our young bodies were pushed to their limits and beyond. We could not enjoy much respite between "Halt! quick march, halt, quick march, left turn, about turn, right turn, halt, quick march! *Lef'*, right *lef' right lef'*, right, *leffff,* squad halt, quick march!" until we were all gasping for breath and soaking with sweat running from the top of our heads, beneath our armpits and between our shoulder blades down our backs. When we were about to drop from exhaustion with our chests heaving, Paddy Malloy would bring the squad to a halt. As we stood to rigid attention, Paddy would run his pace stick down each cadet's spine and then continue the torture until the exertion had drenched each shirt with sweat. We would curse Tony Smit under our breath for inconsiderately continuing to wear a vest under his shirt. This extra layer would absorb the moisture, which extended the drill period until sweat showed up on his shirt as well, prolonging the suffering.

During these sessions, we would spin and twirl around the parade square. Any blunder or oversight in someone's standard of dress would provoke harsh invective and abuse from our tormenter, who would bellow an inch from the face of the unfortunate offender. He noticed the tiniest speck of yolk from a hurried breakfast on my tie on one occasion. His eagle eye missed nothing. He swooped in and blasted straight into my terrified face, "Wrigley! You are supposed to eat your breakfast, not wear it!" We quickly learnt that being addressed was not an invitation to look at him or respond, as he would shout, "Look straight ahead, you miserable ....!" Most of his remarks were hilarious in his heavy Irish brogue, but we avoided inviting his attention with even the hint of a smirk.

Paddy's uniform always drew admiration. He wore a khaki Australian bush hat, one side clipped up, a flat brim, and worn low over fierce green eyes. He always looked immaculate and ready to march straight onto the parade square for the Queen's birthday. His khaki uniform was starched and pressed, with his puttees wrapped meticulously around his calf muscles. He would marshal the squad while continuing to prowl around us with gleaming boots, determinedly crunching the tarmac in military precision.

During one particularly strenuous workout, he marched us at such a rapid pace that the only way I could keep up the tempo was to spread my long legs and swing them outwards by rotating my hips. "Wrigley, you are not part of the 'Folies Bergère'!" was shouted at me after he brought the squad to an abrupt halt.

The first time I attended the gym period, I did not realise that the lithe ginger-headed gym instructor in a white vest and shorts was none other than Paddy Malloy himself, the self-same tyrant who caused us such pain on the drill square. Out of uniform, he was as slender and agile as a gazelle. On cross-country runs, as we puffed along like a herd of elephants, he would float as light as a butterfly beside us, sometimes running backwards to spur us on. He became surprisingly human and likeable during smoke breaks at the gym when he shared jokes.

On one occasion he brought out boxing gloves and invited a volunteer to spar with him. Terry Ryan, who had magnificent broad shoulders, jumped at the opportunity to exact some revenge for the hours of suffering that Paddy had inflicted on us. We shouted encouragement with eager anticipation as our champion stepped into the ring. Surprisingly, Paddy danced nimbly out of reach, quickly ducking each wildly thrown punch. It soon became painfully apparent that Terry was out of his league. Paddy finally floored Terry with a lightning-fast punch, which sent him crashing to the canvas, flat on his back. Following that, each of us had a turn in the ring. Still, having just witnessed Terry's demise, we had a lot more respect for our colour sergeant. We stayed out of arm's reach as much as possible when sparring with him, until he eventually paired us off with another course member. Much later, we learnt that Paddy had been the undefeated, welterweight boxing champion of the British Army.

We attended the Initial Training School (ITS) for the first three months after joining. Here we had lectures on GD (General Duties), service writing, aeronautics, flight instruments, radio, and navigational equipment. These came out of the AP129, a thick yellow manual inherited from the

Royal Air Force, with glossy paper and illustrations. We also had lectures on meteorology, given by Harvey Quaile, the chief civilian Met Officer. He would often read stories from one of his books, which related terrifying accounts and experiences in extreme weather conditions. Since then, the graphic details have instilled a healthy respect for the hazards associated with thunderstorms and icing conditions.

During those first few months, I became desperately unhappy. I did not take well to the process of being broken down and re-moulded into an air force officer and military pilot. I could endure the physical demands of the gym and drill periods but struggled with the seemingly unfair and unreasonable treatment meted out, especially the flimsy excuses used to confine us to barracks or restrict our weekend passes. I found it all demeaning, childish, and unfair. I recall one return to base after a rare weekend pass. As I reached the roundabout that turned onto the Umvuma road, the first sight of the camp made me physically ill, as my stomach knotted up with dreadful despondency. The only thing that kept me from throwing in the towel was my pride. The thought of having to explain to my extended family why I could not hold down my first job was not an option.

Then came the day I strapped into the Percival Provost on my first 'air experience' flight. This aircraft was the piston-engine aircraft used for basic flying training. It was ideal for the job as it needed considerable coordination to make it bend to your will and was demanding enough to weed out those who did not meet the required aptitude. The instructor was F/O (Flying Officer) Dave Thorne, a handsome, fresh-faced pilot with a boxer's nose and charming smile, who had only recently qualified as an instructor. I had seen him at the Officers' Mess and when he cycled, with a group of other single officers, past our marching squad, between the mess and security area. They were always laughing and joking as they pedalled past us. Unlike our flat-topped round air force caps, the commissioned pilots wore theirs carefully shaped by softening them over steaming kettles and folding them over with belts. Their headgear gave them the distinct impression of being 'Battle of Britain' pilots. A casual, contented, and relaxed air was about them, with each cap worn at a jaunty angle. In contrast, our caps had to be worn squarely on our heads, and woe betides any rebellious character who challenged the norm by adjusting his cap angle by a single millimetre. Cherie, Dave's boxer dog, always trotted faithfully alongside this gaggle of cyclists from another realm, and far removed from our lowly state.

This first flight gave us a foretaste of our approaching destiny. It also showed us the light at the end of the tunnel by whetting our appetites. But it sadly highlighted anyone who demonstrated an instant aversion to flying.

I found it a new and exciting experience. An aircraft technician helped me into the parachute harness before I pulled myself up the non-slip walkway at the wing root, by gripping the outside handle. Hunched forward in the cumbersome parachute, and wearing the inner and outer protective helmets, I continued to the cockpit. The soft grey inner helmet incorporated the earphones, which were recessed in padded leather, which stopped them from pressing against one's ears. There was also a grey rubber oxygen mask, which clipped over one's nose and mouth and held the microphone with the on/off switch in front. Behind this inner helmet was the 'pigtail' lead, plugged into a socket for the radio and intercom.

I stepped over the side of the cockpit canopy rail into the hollowed seat pan, shaped to hold the parachute. I sat on the flat surface of the parachute itself, which doubled as a cushion. An assisting ground crew member helped me do up the safety harness and, because of my height, he showed me how to lower the seat fully down and put the rudder pedals far forward.

The instructor completed the before-start checks and then signalled the ground crew for clearance to start the engine. With a blast of white smoke from the exhaust, the whole plane came to life with a roar, and a thrill ran through me as I felt the 550 HP Alvis Leonides 9-cylinder radial engine vibrate through the entire fuselage. It sounded bewildering, and somehow intimate, to hear the flying instructor's friendly voice asking on the intercom through my headphones, "How do you read me?" "Loud and clear Sir," I replied. I was impressed by the professional way he established communication with the control tower to call for taxi clearance; "Good morning tower, this is Foxtrot Two, request taxi for an air experience flight in the GF area". The agreeable answer from the controller was just as clear, "Good morning, Foxtrot Two. You are clear for taxi to runway zero nine. Call ready for take-off." What a great introduction to this incredible world of flying.

The instructor waved for the ground crew to remove the chocks from the main wheels. After a short engine burst against the brakes, he puffed the Provost Aircraft along at idle, as we moved from side to side to see past the high nose.

F/O Thorne's hands darted around the cockpit in new neatly fitting pigskin flying gloves, setting the flaps, and checking the instruments for take-off. We stopped in the engine testing area alongside the grass runway Zero Nine threshold, with the nose facing into the wind. My instructor showed me how to close the cockpit by rotating the handle, which brought the canopy over our heads. I then locked it against the windscreen frame. Holding the control column hard into his stomach, the flying instructor proceeded to do the run-up checks, which required exercising the throttle, pitch control lever, and testing the dual magnetos. The fuselage shuddered with the blast of air, trying to lift the tailplane. I understood then why holding back the control column was necessary to prevent an undesirable occurrence.

When F/O Thorne had completed the pre-take-off

checks, he called the ATC tower, "Foxtrot Two request take off." The tower controller answered, "Foxtrot Two clear take-off, the wind, one-zero-zero, five to ten knots." My instructor replied, "Foxtrot Two, clear take-off."

F/O Thorne taxied the aircraft onto the runway to line up for the take-off run and, as he released the brakes, I was thrilled by the rapid acceleration. I could feel the sharp rudder movements to keep us running straight down the runway, and as the tail lifted, we bounced along in a level-nose attitude, as the grass flashed past. The instructor eased back on the control column at take-off speed, and the aircraft soared effortlessly into the air. As the ground fell away, the rushing sensation of speed ceased. The world around us expanded and seemed to almost come to a standstill. At that moment, I was instantly hooked and knew that this was what I would love! I was going to enjoy this career. Flying was not working; it was living!

2 MIKE HILL – A MAN OF CHARACTER

I first saw Mike Hill at Prince Edward School in Salisbury, where we were both pupils. I next came across him when I attended pilot selection at the Air Force Officers' Mess at New Sarum, but I only got to know him after joining No. 14 PTC on our initial training at Thornhill Air Force Station Gwelo.

He was a year older than me and, on leaving school, had moved to Ndola in Northern Rhodesia to work for Customs and Immigration at the border post with the Belgian Congo. He had a good build and a resemblance to James Dean, the Hollywood film star, with his blond wave curled back and pouting good looks that struck a chord with rebel teenagers of our era. Mike had big feet and never looked comfortable whenever we had to run, but there was an effortless beauty about him in the water. The water flowed like oil over his smooth-toned muscles when he swam. At that time, he was the breaststroke champion of the Federation of Rhodesia and Nyasaland and held the record for the 200-metre distance.

During the time I was in the company of Mike, I never saw him flap about anything. He had an uncanny knack for reading the manual ahead of the lecturer in class, and while the rest of us studied late into the night before an exam, I never saw him prepare for any examination. The rest of us would continually pop into each other's rooms with questions, desperately searching for help on various subjects. Still, whenever I passed Mike's bedroom, he was propped up on his bed reading some cowboy novel, with a cigarette hanging from his lip, and listening to his valve radio with its missing wood casing. The next day, just before the instructor arrived with the exam paper, Mike would wander around from group to group with the usual cigarette dangling from his mouth as he listened to our talking about what we had been studying. He never contributed to the discussions, but I suppose he always gleaned something from what he overheard, and without spending any time on learning, he continued to come in the top three positions in class.

In class, the lecturers would think Mike was not paying any attention because he always sat with his head down, reading ahead of the lecturer. Sometimes, one of them would try to catch him out with a probing question on the subject, but they soon realised this futility as he never failed to come up with the correct answer. I can only assume that Mike must have had a photographic memory, as he could easily remember every card put on the table when he played contract bridge.

Before ITS, we did not have the luxury of having a batman to do our laundry and clean the premises. We had to spit and polish our drill boots and Air Force-issue shoes. We also had to apply Blanco to our belts, puttees, white epaulettes, and cap bands. While the rest of us were always in a rush to get everything crisply folded and in place in time for the inspection, Mike continually proved to be the master of getting by with the minimum effort and fuss. He dispensed with his sheets, pillow, and one blanket, which were left continually on display on his bedside locker and slept happily between the under blanket and mattress. While the rest of us hurried about in a last-minute panic, he would quickly smooth out the under-blanket, lift the pile off his locker, and place it neatly in its position to be ready, long before the inspecting officer arrived. Once we got to the ITS stage, we were moved out of the two dormitories and allocated separate rooms with the luxury of having a few batmen to look after our rooms and laundry.

Mike and I became close friends, but his 'know-all' attitude could sometimes be quite annoying, although I am sure he was right most of the time. Later, his wife told me that his wealth of general knowledge was one thing she admired most about him.

On his 21st birthday, he inherited some money from his grandmother and proceeded to buy a Triumph TR3 sports car and a 3.4-litre Jaguar. He planned to put the Jaguar engine into the Triumph. The new engine stuck out ahead of the radiator and caused a significant problem, and nobody ever came up with a workable solution. I must admit I thought the whole concept ludicrous, as I was sure that the torque would have caused the miniature sports car to do a complete rollover sideways if he ever got it on the road.

Mike was a confident driver, and I once had the dubious thrill of sitting as a passenger and being driven from Bulawayo to Gwelo in his father's Austin Cambridge. Over the next 104 miles, despite some scary bends, the speed never dropped below 100 mph. Mike covered the distance in 1 hour and 5 minutes. He also drove at a ridiculous speed between the two Air Force bases and held the record time for that distance.

We often drove together to Salisbury and other places, and one evening we went to a 'do' at the Hellenic Hall where I took a young lady, the cousin of another course member, Doug Pasea. Doug did not make it through to receiving Wings but later re-joined the RRAF as a navigator and became an air traffic control officer. At this dance, there was quite a crowd with loud rock n roll music, cool 'John Travoltas' and girls with light pink lipstick, heavy mascara, and fluffy soft-coloured tight jerseys, with their skirts flared out by multiple petticoats. As I walked off the dance floor with Pookie, to an exit door onto the closed-in veranda area, I did not notice some lout blocking her way. But Mike, who was right next to us did, and calmly gripped the offender's wrist, and without a word, pulled him aside. I think the grip's recognisable strength was enough to deter any more fuss from the humbled hero.

After receiving our coveted wings and earning the commissioned rank of Pilot Officer, the Air Force posted Brian Strickland, Mike and me to the Jet Fighter Squadron that operated the two-seater Vampire T11 and sporty single-seat

Vampire FB9.

One day, during our OCU training course, Mike and I had to fly a formation sortie in the FB9s. It was still a novelty to carry this out with a fellow course member, which was not allowed before wings. For some reason, I must have been in an extraordinary frame of mind. We were flying in incredibly turbulent conditions, and as I watched Mike struggling to hold a steady position next to me, I could not help laughing into my oxygen mask. This strangely euphoric mood continued as we joined the circuit 500 feet above the ground for a low-level formation break.

The Air Force utilised a spectacular way to join the circuit for an approach and landing. This technique was awe-inspiring when four or more aircraft buzzed in at high-speed overhead the airfield in 'echelon' formation, in preparation for a left or right-hand 'break'. At the appropriate moment, the leader would separate from the others by entering a climbing turn through 180°. He would simultaneously select the dive brakes and throttle back to reduce speed. The remaining pilots would delay their break by one second before following the same path and copying the leader's procedure. They would settle down approximately 400 metres behind each other.

When each pilot levelled off on the downwind leg, he would retract the dive brakes, lower the undercarriage, select the first notch of flap, and apply power to counter the increased drag. The leader and each pilot would level off at circuit height (1,000 feet above ground) on the downwind leg, parallel to the runway. He would continue at circuit height until he reached the point to commence the final 180° curved descent to the runway.

This method's primary purpose was to bring home a group of aircraft and establish them on the final approach to the runway, with a safe distance between each of the following aircraft. It also avoided the danger of flying into the jet wake turbulence made by the plane ahead. Each pilot cut the throttle before starting their descending turn onto finals. He would then control his aircraft's speed reduction by progressively applying more landing flaps, thus eliminating any wake.

I was the lead aircraft for this sortie, and as we levelled off on the downwind leg, the Tower Controller instructed us to go around again as they had noticed some obstruction on the runway. So, I raised the undercarriage lever, which I had just selected to the down position, and added power to hold level flight to extend the circuit.

I reached a point when I was about to fly overhead the field for a further circuit, when ATC said the runway was now clear for a landing. I was then surprised when Mike, who should have followed behind me, calmly said he would go ahead and land before me. I was put out by this, mainly because it was standard procedure to maintain the correct order when flying as a formation. However, Mike decided to do this because he had just reached the point where we usually started the final turn.

Pilots like to do things in a particular sequence and routine, which gives them a sense of security when they follow a familiar chain of events. Things can sometimes go wrong when we alter this usual sequence. This change surprised me since I now had to reselect the undercarriage down again. However, I did this without looking at the lever, and unintentionally selected the dive brake instead. Without making a visual check of the lights to confirm that the three undercarriage wheels were down and locked, I then transmitted "Finals, Three Greens."

The tower controller replied, "Red leader, cleared to land". Although it was one of his duties to visually check each aircraft on the landing approach, to confirm he could see the wheels down, he failed to notice the impending 'wheels-up landing'. Then, just before reaching the threshold, a car waiting to cross the runway, started to drive across. I was about to carry out an overshoot to avoid this danger when the vehicle moved off the runway. Since the runway was now clear, I continued my approach to land. Once I levelled off for the landing, my plane floated right past the control-tower before rolling on smoothly. The only problem was that what touched the tarmac was not the wheels, but the underbelly. By the time I heard the scraping sound, it was too late to go around.

The Senior Air Traffic Control Officer (SATCO) was the person who had driven his vehicle onto the runway when he noticed the FB9 was approaching with its wheels up. He hoped I would abort the landing but did not wish to push his luck too far.

When I later walked into the Squadron, I was given a sympathetic greeting by Mike Saunders, who was then our flight commander, "Welcome to an elite club of senior officers who have also landed 'wheels-up.'"

Shortly after that, the Air Force for some unknown reason, transferred my friend to the Internal Security Squadron that operated the Piston Provost from New Sarum Air Force Base. After that, I lost contact with Mike for some time until I returned from a couple of years in the Fleet Air Arm. We were then once again together on No 1 Squadron, flying the beautiful Hunter FGA9 jet fighter. However, we did meet before this by coincidence when I returned to Rhodesia on two weeks' compassionate leave from where I had been operating off HMS *Victorious* in the Far East. Tish had just recently given birth to a baby boy who had tragically died soon after being born. After two weeks with Tish in Rhodesia, I had to fly back to Singapore via Aden and Bahrain on a Royal Air Force Beverley transport aircraft. As I walked out to climb on board, I was thrilled to see Mike walking out simultaneously. He was on his way to the UK to do his Hunter conversion. After reaching Aden, we spent a day together walking around the shops and bazaars, taking advantage of Aden's status as a duty-free port under British control. We flew on to Bahrain the next day, where we spent one night at the Royal Air Force base there and watched a movie at an open-air theatre under the stars.

While I was in the Fleet Air Arm, Mike married his childhood sweetheart, Judy Hobson, who he had met at Ndola. As a beautiful fourteen-year-old schoolgirl with a cute, upturned nose, full red lips, and waist-length black hair, she was the catch of the town. Even at that age, she was strong-willed and managed to keep Mike firmly under her spell throughout their courtship. But after they got married and returned from their honeymoon, they quickly settled into a Married Quarter's house at Thornhill. On the first Friday back at work, Mike went straight to the pub at the Officers' Mess at the 'knock-off' time of 1330. This practice was such a well-established tradition that he had not considered it important enough to enlighten Judy. She was used to her father's ways, who had a nervous wife and three daughters and was always home before dark. Now, for the first time, Judy was in a strange house, in a different place, and not enjoying being alone after dark, and by the time Mike arrived home, she was furious. He wandered in at ten o'clock that night, in his inimitable laid-back manner, and was most surprised by his reception. But without a word, the new castigated husband, turned on his heels. He eventually returned home well after midnight, in a far worse state of intoxication. This episode quietly established Mike's authority in their marriage, and Judy realised the futility of repeating such a scene.

I believe that Mike had a severe stutter as a child. However, the only thing that I occasionally noticed was the slightest hesitation in his speech when one would expect some degree of excitement. There was just a faint hint of a stutter under the most stressful situations. But through sheer determination, he had managed to bring this under his control, which may have contributed to his casual attitude.

When we were flying on the Hunter Squadron, HQ

scrambled two aircraft. Mike was one of two pilots tasked to carry out a cross-border strike. With envy as well as a degree of vicarious pleasure, I helped him prepare his Fablon-covered 1:1,000,000 scale map, first drawing the track in a grease pencil and then filling in the '2-minute marks' using the Squadron's navigation ruler. At the same time, he quickly donned his flying overalls and calf-length black boots, after fitting the 'G-suit' with its attachment hose dangling out of the slit in the overalls, below his left hip. During the hype, Mike kept his composure. But just as the pilots had completed strapping in, the Squadron Commander called off the flight. My heart sank with disappointment for Mike, but as I shot a glance in his direction, the only hint that he felt any regret was a momentary change when his face flushed a bright red.

Another example of Mike's restraint was when he experienced a hydraulic system failure in the Hunter he was flying, caused by a piece of shrapnel during a rocket attack at the Kutanga Range. He needed to use the emergency system to lower the undercarriage and carry out a high-speed landing. This required fire jeeps to be at the runway side in preparation for a flapless landing. Although the whole episode added a bit of drama to our day, Mike returned to the Squadron building, looking as though nothing out of the ordinary had occurred. But what was even more surprising was that he did not think it even warranted a mention to Judy. She only learned about it that evening, when she overheard some other wives discussing it in the Grog Spot's mess sports pub.

When Mike and I flew the Hunter aircraft on No 1 Squadron, our families became close friends, and we both had young sons born a month apart. They often came out to the home we rented away from married quarters which had an old farmhouse on one hundred acres called Harmony Farm. This was on the far side of a ridge to the north of the airfield, where we constructed a crude cement paddle pool for the two youngsters. They would splash around in it while we sat in the shade of the loquat trees enjoying tea or lunch. We also started doing weight training regularly after work at the Air Force gym at Thornhill. But a year later, Mike, Judy, and Rodney were transferred to Salisbury, when HQ posted Mike to No 7 Squadron to fly the Alouette helicopter. However, we occasionally stayed with them in their newly purchased house, close to the New Sarum Air Force base.

The Air Force sent me to train as a PAI (Pilot Attack Instructor) from the Hunter Squadron. But once qualified, to my disappointment, HQ transferred me to the Training Squadron to become a 'Basic' and 'Advanced' flying instructor. Mike and I did not work together again for a few years. Then the Air Force sent me as an operation assistant to a temporary army base at Lupane. I joined as part of the established JOC (Joint Operations Committee) after ZAPU insurgents infiltrated the area. The JOC also asked for two helicopters to assist the operation. Mike flew one of the helicopters, and a likeable South African pilot, Jerry du Plessis, operated the other. Jerry had a broad and cheeky smile with a mischievous sense of humour.

Also attached to this operation was Alan Savory, a serious and dedicated man who had pioneered and introduced follow-up procedures by means of tracking terrorists' spoor, which became a standard and most effective practice. He also established the principle of rotating pastures for cattle ranching, which proved phenomenally successful. Alan later became one of the youngest members of the RF Party in the Rhodesian Parliament. He was passionate about his beliefs and very intense as he expounded his theories. But to our discredit, we found his ideas boring and somewhat painful to listen to. In the middle of one of his constant discourses, Jerry du Plessis, with a straight face, interrupted him by saying, "I have a powder which I can put on my *veldskoens*. I bet you when I go for a walk tomorrow, you won't be able to follow me." Well, at this outlandish statement, Alan Savory's eyes shot open in stunned confusion and uncertainty. Taking himself very seriously, he was unaccustomed to this kind of childish flippancy. The rest of us had to restrain ourselves from laughing out loud, and later, as we walked through the dark to our tent accommodation, I said to Jerry, "What the ding-dong was that all about?" To which he replied, "I would just walk to my helicopter and get airborne, and my tracks would disappear into thin air."

It was a time in our lives when everything was light-hearted and, while camping out in the 'sticks' with the RAR (Rhodesian African Rifles), we became incredibly spoilt. They operated in style, with a red carpet laid out on the cleared bush in the Officers' Mess tent and smartly kitted-out waiters, serving sumptuously prepared meals with proper crockery and cutlery. One night I pounced on Jerry's back as he rewound his way to his tent along the unlit path. He screamed so loudly in mortal fear, that I could hardly stand while I laughed so much that my sides ached.

Not much happened most of the day, and the only other bit of excitement was caused one night by an accidental discharge of a RAR soldier's weapon. I was already on my stretcher, drifting happily off to sleep, when a loud explosion occurred. The sound of running footsteps followed this up, and for a while, I wondered if we were under attack, but the all-clear quickly followed, and we later heard it had just been an accidental discharge.

During this deployment, Mike had to fly his helicopter to the New Sarum Air Force base to collect some equipment. There was little going on, and in a spontaneous and rebellious moment, I asked if he could route via Thornhill and drop me at home to spend a day with Tish and then collect me on his way back to Lupani. He was happy to indulge me, so I asked the other ops assistants to cover for me. Later that evening, I heard that while I was away, the operations officer at the Lupani JOC, S/Ldr Pete Nicholls, asked where I was. Jerry informed him that I was sick with a bad case of diarrhoea. I had been absent for over six hours that day. But when I appeared for dinner, S/Ldr Nicholls kindly asked if I felt better now, remaining unaware that I had even been AWOL.

A few years later, I followed Mike onto helicopters when HQ posted me to No 7 Squadron. By this time, there was an additional little girl born to each family, only a few months apart. The friendship between our families picked up from where it had left off.

Over the years, I heard more stories about Mike's casual and unruffled approach to his flying. One of these was told to me by Bob Mackie, the flight engineer on a sortie with Mike from the Rushinga base near Mount Darwin. Apparently, after deploying troops near the Mavuradhona mountain range, they ran low on fuel but knew that an RLI unit was looking after a supply of fuel drums on the road to Mukumbura. Mike knew that the Alouette's fuel warning light only guaranteed ten minutes of safe flying time. But when they reached where they expected to find the fuel, they had already flown 18 minutes on the red light. On landing, a trooper informed them that the fuel drums' stock was another five kilometres up the road. Bob could not believe it when Mike calmly restarted the helicopter again. As the flight engineer, he sat on the edge of his seat with his heart in his throat, watching every small clearing in case the engine flamed out.

Another story about Mike was when he went for mountain flying training in the Chimanimani Mountains with Peter Petter-Bowyer. They carried out a power check and inspection of an LZ. But when they had completed a circuit for an approach and landing, the wind had unexpectedly swung around. Luckily for them, this happened over a clear area as there was insufficient reserve power to carry out an overshoot. This limited power left them no choice but to continue with an uncontrolled hard landing in the LZ. As P-B took a deep breath at their narrow escape, he could not believe Mike's casual comment, "I guess we ran out of upward going lever!"

Then, an unimaginable tragedy struck during our son Shane's fourth birthday. Mike was flying as the safety officer during an instrument training exercise. The Squadron Commander, S/Ldr Gordon Nettleton, was the pilot flying on instruments behind a canvas screen that blocked his view outside the aircraft.

At 300 feet AGL, while making a simulated approach onto Runway 33, their helicopter pulled up sharply, rolled over and plunged straight into the ground. The aircraft crash killed both pilots instantly when it smashed into the airfield perimeter fence.

I was at the Squadron when the shocking news reached us. As no one knew where Judy was then, I informed the Station Adjutant that she might be at our son's birthday party in Hatfield.

Following every fatal accident, it was standard procedure for the Station Commander to deliver the tragic news to the wife. And it was always a foreboding moment when an Air Force staff car arrived at your gate. However, in this instance, Tish was happily unaware of the omen and cheerfully went out to meet it. The other wives were more aware of the possible ominous reason for the visit and huddled together in mutual concern. Of course, Judy was utterly unaware that anyone would come looking for her there, but the Station Commander asked sympathetically if Mrs Hill was at our house. So, Tish innocently called for her. The dreadful news instantly shattered Judy's whole life.

The accident happened a few months short of Mike's completing ten years in the Air Force, after which her widow's pension would have increased substantially. In this case, Judy had to seek employment immediately. But as both her children were too young to go to school, and our children were the same age, Tish offered to share a maid to assist in looking after all four of them.

As a military pilot, I have experienced the loss of many friends in flying accidents. Although it has always come as an unexpected shock, Mike's death hit me particularly hard. Not only did we intimately share Judy's grief, but I also found it challenging to come to terms with the fact that this could have happened to him. Mike had always seemed to be immune to things that affected other people. The board of inquiry could never give a plausible reason for the accident, especially after the technical staff sent from France by Aerospatiale claimed that the helicopter had not experienced any mechanical defect.

The only thing that the board of inquiry could substantiate was that Mike had been the flying pilot right up to the moment of impact. Evidence indicated that he had made a desperate attempt to salvage the situation. The position of the yaw pedals and his broken leg and lap strap confirmed this. At one point before the final crash, the engine had cut. The aircraft struck the ground on its belly with the collective fully up.

After that, Mike's accident was to have a detrimental effect on my attitude towards life. Especially when I realised that you could be flying along happily one minute, blissfully unaware that the next moment you could lose control and be killed within seconds.

I also did some silly things in my search for answers, by going to several spirit mediums in a desperate attempt to contact Mike in my search for answers. But I later came to regret dabbling in that sort of thing as I believe it can open one up to supernaturally demonic forces.

Years later, I sadly learnt that Judy had heard a rumour that the Air Force had covered up the real reason for the accident and that I had been party to this, which was, of course, entirely false.

Tish was a great help to Judy over this challenging time, and they would talk for hours each day when Judy came back to collect the children. Over the years, she had many suitors but never remarried, as it would take an extraordinary person to match up to all the exceptional qualities that Mike had.

Judy is still an extremely strong-willed woman living in Zimbabwe. Tragically, their son Rodney, who was four years old when his father died, and inherited Mike's looks and many other attributes, was killed in a motorbike accident at the age of twenty-two.

Toni, who was eighteen months at the time of the accident, has grown into a beautiful woman. She married Stan, a pilot who now flies for Air Emirates. I had the distinct privilege to pay tribute to both Mike and Rodney at their wedding. Stan is one of those rare quiet, and strong men and has continued to support both Judy and his parents.

At the time of writing, Stan and Toni have three beautiful children, and it is plain to see that Judy, her daughter Toni and granddaughter Ronnie-Michelle, all come out of the same mould with their cute, upturned noses.

3 HAWKER HUNTER – FIRST SOLO

Hawker Hunter F6

We thronged the hard-standing area in front of OC Flying's office at Thornhill, anticipating the first two Hawker Hunters' arrival. There was a ripple of excitement when word reached us that the two jet fighters had just flown low-level over New Sarum Air Force base.

The Air Force acquired the iconic World War II Spitfire aircraft not long after the war. They were replaced in 1954 by the Vampire Jet fighters, which had since served faithfully as No 1 Squadron's front-line fighter aircraft, and usually took 25 minutes to fly from New Sarum to Thornhill.

Unbelievably, less than ten minutes after receiving the news, we were surprised by a beautiful, deep, rich resonance signalling the approach of this first pair of Hunters which F/Lts Mike Saunders and Frank Gait-Smith had just ferried from the UK. Within seconds we caught sight of their aesthetically beautiful lines. Flying 400 metres apart in low-level battle-formation, they thundered overhead the field with an ear-shattering roar at just below the speed of sound. For the next five seconds, there were bursts of crackling static in their wake, sounding like leaves swirling all around us.

Enthralled, we watched the aircraft swoop back into view over the field, with their long fuselages and swept-back wings silhouetted against a halcyon sky. The beautiful symmetry of these aircraft took my breath away. Frank slid his plane gracefully into close formation as they joined the circuit from the 'Initial' reporting point north of Runway 13. Flying parallel to the runway at 200 feet AGL, they snapped into a low-level formation break from abeam the spellbound onlookers. The gap opened as each aircraft curved into a left-hand climbing turn onto the downwind leg. Each plane continued a curved descending approach to touch down with a puff of white smoke. The drag chutes then popped out and shimmered behind each one, slowing them down to taxi speed. As each aircraft turned off at the end of the runway, the pilots jettisoned their chutes like spent balloons.

The two aircraft were taxied into the dispersal area, right before us and marshalled onto their parking spots. The pilots synchronised the engine shut down, leaving a wisp of lingering exhaust fumes curling out of their long jet pipes. The Hunter's tailpipes were so high that smaller pilots and technicians could not see down its length to check the Rolls Royce Avon 204 engines' turbine blades. I was amazed by their overall length compared to the squat, bulbous Vampires with short downward-pointing jet pipes.

These new arrivals shone with a glossy coated camouflage brown and green upper surface, with blue-grey underbelly and wings. As I walked around this exquisite aircraft, I ran my fingers over the smooth, enamelled fuselage. I could hardly believe that I would be attending the next conversion course in the UK a month later.

It was now the beginning of 1963, and I had been flying Vampires on No 1 Squadron for nine months, after having been presented with my coveted wings and commissioned to the rank of Pilot Officer.

Just short of my 21st birthday, I was the youngest pilot on the Squadron and felt blessed to be part of an elite few, leading the life of a jet fighter pilot. The next youngest on the Squadron was five years older than me, and most were already married with children. In their mid-twenties, they all seemed so mature and experienced to me, and I admired these wild quixotic bunch of seasoned pilots.

When I joined, the Air Force had only recently posted over half of the pilots to the Squadron, by transferring them from the Internal Security Squadron, which operated the Piston Provost aircraft out of New Sarum Air Force base. Mike Hill, Brian Strickland, and I were all posted to No 1 Squadron after 'Wings' to undergo OCU training, in which we learnt how to use the aircraft as a weapons delivery platform.

I was in heaven each day I went to work. It was such a unique lifestyle and not like real work. I was so proud to wear the wings on my chest, which signified I was now a qualified Air Force pilot. The single sliver of rank worn on the shoulder epaulettes, or Air Force jacket sleeves, indicated I was a commissioned officer in the Royal Rhodesian Air Force (RRAF). This career was life! Every morning I would cycle happily from single quarters and wind my way through married quarters to have breakfast at the Officers' Mess. After that, I usually joined a gaggle of other young unmarried officers cycling to work along an avenue of jacaranda and flamboyant trees. After passing in front of station headquarters, we entered the security area under the raised boom to report to the aircrew briefing room at 0630. This briefing took place in a prefabricated building, to the left and opposite, the Ground Training School buildings. Harvey Quaile would give a met report and F/Lt Don Brenchley, the chief navigation officer on No 5 Squadron, called out the exact time to synchronise our identical aircrew issue watches. The pilots and navigators continued to their various squadrons when the briefing had finished. They walked or cycled down more avenues of exotic flowering trees and past the equipment stores and other offices constructed during World War II. The Air Force continued to paint each building with cream-coloured corrugated-iron walls and maroon roofs. The interior walls were timber-framed and covered by soft white ceiling board sections. Each hangar had identical squadron buildings attached, which incorporated various offices and pilots' crew rooms. The atmosphere was always benign and friendly in our crew room, as pilots gathered for coffee. Adhering to tradition, the three JPs (junior pilots) prepared and delivered these beverages to the other pilots.

These senior pilots, whom we referred to as the 'heavies,' usually sat around reading newspapers or challenging each other to games of darts. It always amazed me how many individual styles there were to throw a dart, but despite that, most of the pilots displayed surprising accuracy.

Flt Lt McClurg was the Squadron PAI conducting the course. After coffee, he would inform us of the day's flying programme. I enjoyed each new exercise, and particularly enjoyed delivering weapons at the Kutanga Bombing Range, near Que Que. Here we fired three-inch rockets, 20 mm cannon rounds, or dropped small 25-pound practice bombs from high dive or low level. Once we had mastered the rudiments in the T11 trainer, we progressed onto the sporty single-seat FB9.

Peter McClurg then taught us the rudiments of air-to-air gunnery, which we flew in a very academic way. The target aircraft held a steady height and speed, while the attacking aircraft positioned a few thousand feet higher, slightly behind and 1,000 metres offset to one side of the target. The extra height allowed the attacking aircraft's speed to increase in a descending turn to close in with the target aircraft. At 100 metres behind, we would break off the attack by reversing the turn and climbing back to a perch position on the opposite side.

A camera mounted on top of the gunsight was activated whenever we pulled the trigger. The film was then examined to determine whether we would have shot down the target aircraft. We could calculate the frames' percentage when the sight's dot was on the target aircraft. These would only count as hits when the distance between the inside tips of the diamonds' ring circling the central pipper, matched the target's wingspan. We controlled the outside diamond width by twisting the throttle grip.

During the advance training on the Vampire T11 before 'Wings', formation flying was only allowed above 10,000 feet AMSL. However, on OCU, when we practised with one of the Squadron pilots flying the lead aircraft, this restriction fell away. On one training sortie, I went up with Frank Gait-Smith, an unusual and unfathomable character who had been a Rhodesian boxing champion and played hockey and football for the country as well. At that time, he was still an extremely active sportsman and would run from one event to the next. He often participated in three different sports sessions in the afternoon and then played a squash game after dark. He was a good-looking man with a typical boxer's nose and a reputation of being unpredictable and cranky.

On this sortie with Frank, he flew as the formation leader. During the entire exercise, I battled to hold a steady formation position. We stayed at the uncomfortably low height of only 200 feet above ground level the whole time. Frank was not concerned by my limited experience as he yanked his aircraft about in steeply banked turns. I struggled to hold a steady position flying the FB9 at a low level and high speed. I had never been so terrified. Either Frank was oblivious to my terror or thought he needed to put this wimp in his place. He continued to pull high G turns one way and then the other and sandwiched me between his aircraft and the ground when he turned towards me. I became so tense that my heart raced, and my thighs began to twitch uncontrollably with sheer terror. Everything in me screamed to pull away, but I dared not, for fear of flying into the ground. Once back at base, it never occurred to me to make a fuss about it. Still, considering my limited experience, it was a dangerous and irresponsible thing to do.

Mike Saunders, Rob Gaunt, Gordon Wright, and Frank Gait-Smith were the first four pilots sent on the Hunter conversion course at RAF Chivenor, in North Devon. They caught the train from London to Barnstable, and on the journey consumed an excessive amount of Double Diamond beer in the saloon coach. When the train stopped at Exeter, they nipped off the train to continue at the station bar. Busy with their drinks, they were unperturbed when the train started pulling out backwards from the station. They mistakenly assumed it was only shunting back and forth and patiently waited for its reappearance. By the time they realised their error, the train was well on its way without them. In desperation, they phoned the RAF station duty officer, who sent a staff car to fetch them. When they finally arrived at Chivenor, it was already late. However, this did not deter them from heading straight to the Officers' Mess bar to continue where they left off. When the pub closed, it dawned on them that they had not eaten the whole day. Now famished, they decided to break into the mess kitchen for some food. Mike felt it his duty to contact the Station Adjutant and report their safe arrival the following day. But the curt response was, "Don't worry, the whole Station already knows you're here!"

I joined the third group of pilots sent to Chivenor and left Salisbury Airport on a hot and muggy February night. It was my first flight on a commercial airline, and I boarded the sleek BOAC four jet-engine Comet, wearing a white shirt with a sports jacket and tie. On our refuelling stopover in Rome, I was surprised by how cold it was, but this turned out to be a mild foretaste of what to expect in London.

Descending through the low cloud on the final approach to Heathrow Airport, it was my first sight of both the UK and snow. A shroud of white covered everything in view, the roads, leafless trees, hedges, roofs, and fields. It was a most unexpected sight, and I was totally unprepared for the coldest winter they had experienced in the last 50 years. When I stepped out of the plane, my ears, nose, and hands suffered the freezing onslaught of the most unbearable, extreme cold I could ever have imagined.

In London, we reported to Rhodesia House to collect our train tickets. While there, we overlooked the lions around Nelson's Column and the myriad pigeons at Trafalgar Square. Later, we enjoyed our first experience of an English pub. I had a half-pint of bitters with a steak and kidney pie.

The outside temperature was 40 degrees Fahrenheit below freezing. Since England reputedly never expects or prepares for such low temperatures, most exposed water pipes remain unclad and frozen solid. It was equally cold when we reached Barnstable. The Chivenor officers' single quarters were black wooden shacks left over from World War II, with difficult-to-operate charcoal heaters. Instead of accommodating us there, the station commander decided to book us into the Barnstable Bell Hotel.

Staying in the hotel was a real bonus. I shared a room with Cyril White, a veteran of the Malayan Campaign, before he joined the RRAF, on a short service commission. His next move had been to leave Rhodesia and join the RAF. He spent a few years flying Venoms and Hunters there before re-enlisting into the RRAF. A quiet, reliable person of few words, he could drink gallons of bitter with no ill effect. Each night I accompanied him on a pub crawl where he would stand leaning against the bar, cradling a beer mug against his chest.

He would savour each drink with quiet contentment, and I could barely manage to keep pace with him. I would have a half-pint each time he was ready for another round.

A revelation was to see scores of young ladies going unescorted to a big dance hall nearby. In Rhodesia, no girl would venture from home unless specifically invited out on a date, on which she would be treated with the utmost courtesy and never expected to pay for anything. This nightly gathering of young English roses at the dance hall was a bonanza. The only challenge was due to the doors being closed at 2130. The manager closed them to block an influx of drunks on their way home from the pubs, which closed simultaneously. My 'mission impossible' was to get Cyril out of the pub in time to reach the dance hall with the assembled flock of friendly lasses.

The RAF pilots that we had met at Chivenor, warned us that the local girls were very antisocial. But one evening at the dance hall, I recall looking across the packed floor to see each of our four pilots up close and friendly with some rosy-cheeked damsel. On the other hand, a bunch of sullen-looking RAF pilots stood glumly against the surrounding walls nursing their beers.

I enjoyed this wartime base's atmosphere, where we had lunch and a high tea at the mess. In Rhodesia's tropical climate, drinking soup while steam was coming off it carried a risk of inflicting a scalded pallet. But it took me a while to realise that I did not have to wait until the soup stopped steaming before I could start to have some.

We were allocated ten hours in the simulator on the Hunter conversion course and about four hours on the two-seat Hunter T7 training aircraft. The T7 was heavier than the single-seat F6 and had a less powerful Rolls Royce engine, producing little over 7,000 pounds of static thrust. This power contrasted with the single-seat version, which delivered over

10,000 pounds.

When it came to my first solo on the Hunter F6, nothing prepared me for the exhilaration I was about to experience. When cleared for take-off, I applied full throttle and released the brakes. After a momentary pause before the engine bleed valves closed, it felt like a standard gentle increase in power. However, I was astonished by the unanticipated surge of thrust that kicked in. It felt as though somebody had lit a rocket behind me. The powerful force pinned me against the headrest of the Martin-Baker ejector seat and the sudden acceleration took my breath away. It was like being on the back of a bolting racehorse as I held on for dear life. Each step of the way, I was behind the aircraft until well above 10,000 feet. I was late to raise the nose wheel off the ground, late to lift off, late to raise the undercarriage, and late to retract the take-off flap. I even overshot the maximum climbing speed limitation of 250 knots below 10,000 feet and nearly missed my level-off altitude of 20,000 feet. I could not believe how far the throttle had to be retarded to maintain a reasonable Mach number for general handling practice. On returning to base, I carried out a few touch-and-go landings. Each time, as the aircraft sped down the runway and I applied full power for the next take-off, there was that deceptive momentary delay, before the engine RPM whizzed up to maximum revs and gave me that exhilarating kick from behind.

The whole time, I went about in a state of stunned elation. There were no words to articulate the feeling, and as no one else made any comment on their solo flights, I assumed I was the only one to have enjoyed such a thrilling and enthralling experience. This flight stimulated me for the rest of that day and well into the night.

In the end, 12 Hunters were ferried out from the UK via Khartoum, Nairobi, and Blantyre to Thornhill. The RAF *fundis* doubted that all 12 Hunters would be delivered intact.

Nevertheless, our ferry pilots successfully achieved this.

Frank Gait-Smith related a problem he had experienced with the rock-hard ejector seat cushion during the long ferry flight. After Khartoum, he came up with the bright idea of folding his G-suit and placing it on top of the seat pan. When his buttocks began to ache, he would press the test button to fill the G-suit with compressed air. His solution worked marvellously to ease the pain. However, when he came in for the break into the circuit at Nairobi Airport, he forgot to switch it off, and the G-force pumped the suit up like a balloon. This inflation pinned him hard up against the canopy. He struggled to switch it off for some time, but much to his relief, eventually managed to do it.

As I was still relatively inexperienced, the flight commander informed me that I would not be participating as one of the ferry pilots. But as I was so in love with the Hunter and grateful to be one of the privileged few Hawker Hunter pioneers on the Squadron, this did not bother me.

Later that same year, I was honoured to fly in the No 4 position in a formation of all 12 Hunters, in the shape of a 'diamond twelve'. I was positioned directly behind S/Ldr Mike Saunders, who the Air Force had appointed as our new Squadron Commander. It was the first and only occasion that all twelve Hunters were up simultaneously. This event occurred around August 1963, and it would be ten years before one aircraft crashed, when Flt Lt Al Bruce had to eject on short finals into Bulawayo.

4 JOINING THE FLEET AIR ARM

With other students on the FGA11

It was a strange chain of events that caused me to leave behind a life of happiness for me in the Royal Rhodesian Air Force, to join the Fleet Air Arm. I had never excelled on the sports field, not for lack of trying, but luckily, I took to flying like a duck to water. I was fortunate enough to have a certain amount of natural ability, and my RRAF flight-assessment reports described me as having 'dash'. The Air Force usually channelled pilots with this trait onto jet fighter squadrons.

After receiving my wings in Rhodesia, the Air Force assigned me to fly fast jet fighters and managed to fulfil my ambitions surprisingly quickly. I found myself posted to No 1 Squadron in 1962. It was due to be re-armed with twelve brand new Hawker Hunter FGA9 ground attack/fighter jets and I could not have been more content with life. These were heady days. I was the youngest pilot on the Squadron among many wild and carefree young men as my role models.

An unexpected turn of events arose in 1963, with the

Federation of Rhodesia and Nyasaland's impending break-up. The terms of the separation of these three countries gave military personnel the option to remain where they were, join one of the British armed forces or transfer to either Northern Rhodesia or Nyasaland, which would soon gain independence from Great Britain.

Still at an impressionable age, I was egged on by the bravado of the Squadron's senior pilots who fantasised about life in the Royal Navy or RAF. The discussions sounded exciting as they painted exhilarating and rosy flying scenarios in and out of exotic locations. These discussions convinced me that most of the Squadron pilots would soon be heading off to a life of adventure and visits to far-flung ports worldwide in the Fleet Air Arm.

At this juncture, I met Tish, the sister of Dave Thorne, in her role as a bridesmaid at his wedding. From the time he was my flying instructor, Dave's manner and charm had always fascinated me, and I was now a guest at his wedding. His marriage to Rosalie took place in Gwelo, with the reception at the Thornhill Officers Mess. We were greeted by Dave's family at the front entrance. Tish was stunning, with a beautiful smile. In my macho Air Force style, I asked, "Do we get a kiss from the bridesmaids?" Still smiling, she responded, "Certainly not!" Undeterred, I felt an instant spark of attraction between us, which was soon confirmed. Every time I looked at the main table, she would be looking my way and continually bestowing her lovely smile on me. When most guests had left the formal reception, officers and their wives drifted across the lawn for an impromptu party at the Grog Spot. The single officers invited some of the guests along, and I persuaded Tish to join us. We immediately hit it off, and on the dance floor, I enjoyed her bubbly and sparkling charm. Later, we ended up kissing in the back of Dave Currie's car. He, as the groomsman, waited patiently to take Tish and her younger sister Heather to where the bridesmaids were staying that night.

The following week it took a few abandoned attempts to phone Tish at her workplace in Salisbury. I first needed to still my beating heart and stop hyperventilating long enough to dial her number. Every time it started to ring, I lost my nerve and hung up. But after several aborted attempts, I was able to steel myself long enough to persist. I was relieved to recognise Tish's delighted response on the other end of the phone. Our relationship quickly blossomed into a romance, and I travelled 180 miles most weekends to see her in Salisbury. When I left my car with her, Rob Gaunt made a prophetic and accurate pronouncement over my bachelorhood when he teased me with, "doomed!"

Our love deepened despite the arrangements for my transfer to the Fleet Air Arm, which made it a bitter-sweet experience. It broke my heart to leave Tish behind, but on the other hand, I basked in the drama and envy of the other pilots. But at the end of the day, when push came to shove, everyone except Frank Gait-Smith, copped out.

In their defence, most of the other pilots on the Squadron were married, and their wives had other ideas. Only the two of us took up the offer of a transfer to the Fleet Air Arm. Then, at the last minute, Frank changed his mind and decided to join the Royal Air Force instead. As the chain of events gathered pace, I enjoyed the envy of the other pilots, who vicariously lived each step of the way. But as time went by, there was no possible way of my pulling out without upsetting many people, especially after attending the interviews and receiving the letter of acceptance from the British High Commissioner. The Commissioner's office quickly booked my trip to the UK, and I found myself being swept along by a seemingly unstoppable train of events. My fate was now firmly in the hands of the British High Commission and Royal Navy.

On 14 January 1964, I left Salisbury for Cape Town by

train. Tish came with me as far as Gatooma, where her father, who was then the personnel manager of Rio Tinto, at the nearby Eiffel Flats Mine, came to collect her but kept discreetly out of sight. Tish looked pale but put on a brave face, having made plans to visit me halfway through the following year. She stood at the end of the platform and continued to wave until the train was out of sight. As the train picked up its pace, I hung tearfully out of the window, waving goodbye.

Following a breakdown in the Karoo, a coach arrived to collect the *Edinburgh Castle* passengers. On the trip we had to endure a terrifying ride, plunging down Du Toit's Kloof at breakneck speed, before joining the vessel waiting to depart from Cape Town's Table Bay docks. As a Royal Naval Sub Lieutenant, they had booked me in as a first-class passenger.

Halfway through the two-week cruise to Southampton, I was so homesick and missing Tish, that I proposed to her by telegram.

After a month in the UK of not seeing the sun, the reality of my situation hit me. For the first time in my life, I felt a debilitating blanket of depression overcome me. The sinking feeling encompassed me as I looked out of the classroom window, onto the grey, rain-soaked parade square at Pompey Barracks. I realised I was like a fish out of water. The whole environment suddenly lost its appeal and became foreign to me. Here, I found myself among a bunch of strangers who could not possibly empathise, or understand, how I felt. I was attending an orientation course at Portsmouth to introduce direct entry officers into the Navy. With me in this group was an ex-RAF pilot, who had tried his hand in commerce and industry, a few schoolteachers, plus the only ones I warmed to, a couple of officers from Trinidad and Tobago's small Caribbean navy.

Having been told to share a 'cabin' with one of the teachers, I immediately pictured a log cabin. But soon discovered it was a room in a substantial brick-built double-storey building. This building also held the 'wardroom', corresponding to the Officers' Mess in the Air Force.

I found I needed to bathe every night to get myself warm before bed, but my roommate had not had a single bath, after ten days. I was beginning to think this a bit strange until I overheard him derisively telling one of the other officers in the bar, "You know he baths every night!"

Before this, I had never experienced such anguish. Until that moment, I had led a carefree life, brimming with overconfidence and the arrogance of youth. Now I felt confused and disorientated. I quickly lost all sense of identity or where I belonged and was incredibly homesick. I thought how good it would be to find it all just a bad dream, and that I was still in the RRAF.

The Navy wardroom was the equivalent of the Air Force officer's mess. But when I went for a drink or a meal, I found the atmosphere unbearable. The thinly disguised obsequious behaviour of young officers towards their seniors, turned my stomach. They were so spellbound by every word uttered by the higher-ranking officers, which they interspersed with bursts of raucous laughter.

After the course was completed, I joined up with young pilots undergoing conversion and operational training on the Hunter aircraft at RNS Brawdy in South Wales. A few of them were even 'General List' career officers, who had volunteered to move from their previous qualifications as navigation, engineer, or armament officers. They had all received their commissions to Sub Lieutenant's rank before undergoing initial flying training on the Jet Provost, with the RAF.

My flying was the one thing that kept me going, especially when I started to receive glowing assessments. But, after the more leisurely rhythm of the Rhodesian Air Force flying programme, it took me a while to adjust to the frenetic pace of the Fleet Air Arm, which ran like clockwork. But once used to it, I came into my own, earning the respect of both instructors and students alike. The instructors were particularly astounded after I broke all previous records at the rocket range at St Brides Bay. This bay was close to the city of St David and used large, floating buoys as targets. On one sortie, in a 28-knot crosswind, I destroyed all five of these buoys by direct hits from the two-inch rockets, which shattered them all. Before that, no weapon fired had ever hit one of them. I heard later that the RNAS Brawdy had added my name to the board which displayed me as the top student for 1964.

Another relief from my remorseless misery was when I drove to London to visit my sister over long weekends and short periods of leave. At that time, she was a trainee nurse at the Royal Masonic Hospital, and I found solace in the terrace house in Chiswick, that she shared with other nursing friends.

The correspondence with Tish was the one remaining connection to my previous life, which I had so recently left behind. She continued to offer hope of a future life together and booked a flight to Le Bourget airport in Paris, where I agreed to meet her in July. Fortunately, this coincided with the two-week recesses taken by all Navy personnel based in the UK. Otherwise, I am unsure if I would have dared to request special leave from the pressurised flying programme. By then, it had been seven long months since I had joined the Fleet Air Arm.

At the same time, my cousin Romeo visited the UK and stayed with my sister in London. He kindly offered to accompany me to the airport and drive my car back. But both of us, still young and naive, miscalculated how long it would take to get from Chiswick to Gatwick Airport, and I missed the flight. Thankfully, when I got to the check-in counter, an ex-Rhodesian in the back office recognised my accent and immediately came to my rescue by booking me onto the next flight.

When I disembarked in France, I found the bustling scene at Le Bourget Airport a complete culture shock. This buzz of activity took me way out of my comfort zone. Still inexperienced in the ways of the world, a young con artist saw me coming and ripped me off. He had some hard-luck story and being unfamiliar with the French franc and feeling sorry for him, I handed him some high-denomination notes.

Arriving over an hour late, I wondered how I would ever find Tish in this mass of humanity. Then miraculously, she appeared, pushing her trolley towards me. Her beauty and flowing blonde hair overwhelmed me with excitement and relief. Oblivious of anything else, I just held onto her close, and it was minutes before I could let her go. I had forgotten just how beautiful Tish was, and feeling weak in the knees, I just had to sit down. As we sat there, I could not stop admiring the fine dusting of freckles on the bridge of her nose and cheekbones. I was also impressed by the confident way she hailed a taxi and spoke French, asking the driver to take us to some small Parisian hotel.

We drove back to the naval base on our return to the UK, where I booked Tish into a small bed-and-breakfast cottage, perched on top of the cliffs overlooking St Brides Bay.

Later, when I took her to the wardroom, her slender figure, and tanned legs, showed off in her tight mini skirt, was an immediate hit with the Naval officers. The same ones with whom I had previously had nothing in common, now swarmed around her while she perched prettily on a barstool, cheerfully basking in the admiring attention and graciously accepted the continuously proffered stream of drinks. Excluded from her circle of besotted fans, all I could do was watch with pride in the happy knowledge that her heart belonged to me.

Nevertheless, I was not the same person who had so confidently waved goodbye to her from the train window as it pulled out of Gatooma station, less than a year before. The depression I had carried around like a suffocating blanket had left me a broken and shattered shadow of my former self. Having lost all sense of identity and self-confidence, I could hardly remember the person I had been. But it had been needless to worry that Tish would not be able to relate to the shell of a person I had now become.

Soon after that, Tish went to stay with my sister for a few weeks before I transferred to RNAS Yeovilton, in Somerset. Here I was to begin a conversion onto the Sea Vixen night-fighter, which was an aircraft operated by a pilot, and an observer who was responsible for navigation and control of the airborne radar. The Navy had given me an alternative to train on the single-seater twin-engine Scimitar jet fighter at RNS Lossiemouth, near Aberdeen, in the north of Scotland. But although the aircraft was akin to the Hunter and more to my liking, I chose Somerset because it was much closer to London. After a few weeks there, I managed to find the perfect little flat on the top floor of a delightful old stone farmhouse, not far from the airbase. The owner, Mrs Cobden, was a widow in her seventies, who we both came to love dearly.

On 12 August 1964, we were married in London to coincide with the next two-week recess. Those who celebrated our wedding with us were my sister, Antoinette, and her friends, my cousin Romeo, who had just immigrated to the UK and Alec Roughead, who had been on my cadet training course in the RRAF and was now on leave. We jetted off on our honeymoon for two weeks to Rapallo, on the Italian Riviera. While there, I suffered sunstroke, having been out of the sun for so long, and I came second in a singing competition on the beach singing 'Wooden Heart'. A memorable time.

On our return, we moved into our beautiful upstairs flat in the tiny, rural hamlet of Coat, near Martock in Somerset. Here we would spend four glorious months of married life.

Tish was a stabilising influence in my life and helped me regain self-confidence and self-esteem. During the course, I quickly took to the powerful Sea Vixen aircraft. Its twin booms, high tailplane, and offset cockpit looked formidable. There was no two-seat version, so instruction began in basic first-generation simulators. When I had completed ten hours in the simulator, the actual aircraft's initial flight was a brand-new experience.

Standing next to the Sea Vixen, I felt utterly dwarfed by the sheer enormity of the aircraft. Its maximum take-off weight was over 20 tons, and its two Rolls-Royce Avon 208 jet engines produced a total of 22,500 pounds of static thrust.

The prototype Sea Vixen was way ahead of its time. But governmental prevarication had delayed its release into service, and although the engines could accelerate the aircraft to supersonic flight, the thick-cambered wings were a limiting factor. From the time of their inauguration, there were heavy aircrew losses. From 1962 to 1970, thirty accidents accounted for the loss of fifty-one aircrew members. By the time I joined the Squadron, the pilot's operational life expectancy flying off aircraft carriers, was only two years.

However, I enjoyed the Sea Vixen's performance and capabilities despite its reputation. The large area of the control surfaces and wings enhanced its performance. I was especially impressed to find it possible to maintain a steady 45-degree bank turn while maintaining an airspeed of Mach 0.85 at 45,000 feet and the ability to perform a perfectly controlled loop above 30,000 feet.

Because of the massive control surface area, the aircraft designers incorporated an artificially induced resistance to the hydraulic control column, making it heavier as the airspeed increased. The manufacturers named this system 'Force Feel', which helped prevent aircraft overstress. However, it also gave me an added advantage in aerial combat exercises as I had more strength than most other pilots. By the end of the course, this contributed towards my attaining an unbeaten record in aerial combat. The enjoyment of this reputation motivated me to analyse further and improve my *modus operandi*. Each night I would describe to Tish the different tactics I had executed by re-enacting the manoeuvres with our toothbrushes. The other advantage and blessing I had was to be long-sighted, which made it possible to spot the opposing aircraft long before they saw me.

We would set off as a pair in close formation for aerial combat exercises. When the two aircraft got to a pre-selected area, the leader called for the split. Each pilot would turn his plane 45 degrees away from the other and start timing on his stopwatch. He would then hold the new heading and maintain altitude and speed for one minute before turning back 90 degrees towards one another. Each opposing pilot had to keep his height and airspeed until he saw the other aircraft.

Most other pilots automatically went into a climbing turn, which diminished their performance. This initial climb would inevitably end up opposite one another in a 'Circle of Joy'.

They would then cut the corner by first going into a 'high speed' dive and then pulling up into a 'low-speed yo-yo.' Invariably no one managed to gain much ground on the other by using this system.

On my introduction to this exercise, the programme sent me up in opposition to a visiting 'ace' war veteran, with combat experience and already a legend, by the name of Commander Bolt. Before walking out to my aircraft, I asked a friendly and noticeably well-bred instructor from the British upper class, and the current British bobsleigh champion, for some tips. He suggested I mirror whatever manoeuvre the commander carried out. So, when we got into the circle of joy, every time he dived, I dived. When he climbed, I climbed.

Then, by some miracle, I managed to pull a little bit harder than he did, and unexpectedly ended up on his tail. When I called out, "Murder, murder", he was amazed and mortified to be beaten by a raw student, who was still busy learning to fly the Sea Vixen. Once the story got out, the other students and instructors on the No 766 Squadron, were stunned.

Having got such a kick out of this newly discovered talent, I did some additional research on obtaining maximum performance on the aircraft. Then, once I understood a few principles, I devised a technique that worked well for me, which commenced when I caught sight of the opponent.

I would immediately dive to about 4,000 feet below it. In this way, I was able to keep my opponent visual as a silhouette against the sky. On most occasions, I commenced this dive before the adversary had even caught sight of my aircraft with its camouflage finish concealing it from above. No one expected the 'enemy aircraft' to come from below. I accelerated to the optimum speed of 460 knots in the dive, making it possible to pull the tightest turn at the maximum allowable 6G, and just below the high-speed stall. By judging the precise moment, I would then pull up into a loop, and if it were well-judged, I did a roll off the top and ended up directly behind the opposing aircraft. Inevitably, I would zoom up from below the horizon without ever being spotted. The first indication of me behind them was when I called, "Murder, murder!" to indicate I could fire the 'Red Top' air-to-air missile. The other pilots could not figure out how I could continually outmanoeuvre them without ever being seen. I had no intention of enlightening them either! After a year in the Fleet Air Arm on Sea Vixens, I remained unbeaten in formal and informal aerial combat.

The manufacturers designed the Sea Vixen as an all-weather/night fighter aircraft to comply with the requested specifications. During the training Squadron's operational conversion course, they devoted one whole month to radar-controlled interceptions at night. During that month, I flew 59 hours, of which 29 hours were night flying. This protracted flying at night was a bit of a shock to my system. Before that, in the RRAF, I had only flown the minimum requirement of one sortie per month to maintain my night flying currency. Invariably we would only fly out on a bright moonlit night.

Each night we flew west of the airfield off the Devon coast to carry out the exercise over the sea. To make things unusually challenging, we had to fly in solid cloud cover from 200 feet AGL to 30,000 feet over that whole month.

Each night, two aircraft would take off in formation, with one of the instructors flying the lead aircraft. This instructor would fly 1,500 feet above the sea as the target aircraft. The student would then separate from him as the intercepting aircraft at only 1,000 feet above the water. Once the Sea Vixens split apart, they came under the control and direction of an airborne radar operator, who sat in front of a consul and radar screen in one of the ugly Fairey Gannet aircraft, powered by two counter-rotating turboprops. The bulging lower belly of this strange-looking aircraft housed the radar scanner. The same controller would vector the target and intercept Sea Vixens on two radio frequencies. As the fuel consumption in jet aircraft is excessive at low altitudes, both aircraft flew at low speed and one engine shut down most of the time.

The controller gave the intercepting aircraft headings and speeds to fly and instructed it when to commence a maximum banked turn. These interceptions were carried out from all directions, even head-on, to bring the attacking aircraft within a missile firing range of one nautical mile behind the target. The workload for the attacking pilot became intense during the final interception turn. While flying on instruments, he had to re-light the shut-down engine by accelerating the aircraft from the loitering speed of 220 to over 400 knots, which increased the ram effect into the engine's intake to speed up the rotation of the windmilling jet engine. The pilot then pressed the relight button on the throttle's side, while simultaneously opening the high-pressure fuel cock. Then once the pilot relit the second engine, both throttles had to be extended to maximum power to accelerate to Mach 0.85, while turning the aircraft in a 60-degree bank turn.

Possibly due to fatigue catching up on me after continual late-night flights, and still flying during the day, I suffered from vertigo on each sortie. Each night, I had to fly in the cloud at 1,000 feet above sea level. The snappy changes to the bank, in response to the controller's instructions, with no stabilising visual reference to the natural horizon, accentuated my debilitation. When I occasionally popped out of the low cloud, the sight of a lighthouse on the coast further exacerbated my disorientation, as its rotating beacon seemed to sweep vertically.

Night after night, my senses would tell me I was flying 90 degrees to the horizon, or even upside down. I was so terrified to lose any height and plunge into the sea or gain 500 feet and collide with the target aircraft, that the slightest flicker of the VSI (vertical speed indicator) panicked me to make an immediate correction to the nose attitude on the AH (artificial horizon). Throughout this unpleasant experience, my pride prevented me from disclosing my struggle. Using his radar to vector me on the interception's final stages, my observer in the Sea Vixen was blissfully unaware of our precarious situation.

At the end of the Sea Vixen course, we did some 'touch and goes' on HMS Ark Royal's aircraft carrier in the English Channel. Our first catapult launch training was from a concrete ramp at Bedford. However, this did not have the advantage of being launched off a ship, steaming at 20 knots into the prevailing wind during carrier operations, which creates a headwind for the launch and recovery. Without any headwind, the launches from the ramp were a scary and startling experience. The steam-driven catapult impelled the twenty-ton aircraft at 4G, to accelerate it from a standstill to 140 knots, over only 50 metres.

Before the launch, I taxied the aircraft onto the ramp. Once hooked up by a steel cable harness to the steam-driven shuttle, I selected the take-off flap and checked the two white 'dolls-eyes' to confirm that the wings were in the down and locked position. I then applied full throttle to the two Rolls Royce Avon 208 engines to utilise their combined 22,500 pounds of static thrust. Once the take-off checks were complete, with the heel of my left hand behind the throttles and both elbows forced into my stomach to avoid pulling back on the control column, I gave a quick nod of my head to signal, "Ready for launch" and then pushed my 'bone-dome' back against the headrest.

The kick from behind and the headlong rush were so exhilarating, they took my breath away. But before I had time to savour the moment, I had to give full attention, rotate the nose to the take-off attitude and select the undercarriage up. I then repositioned the aircraft for a landing back on the runway for another go. I found this an experience that left me on cloud nine for the rest of that day.

While at Bedford, I took the opportunity to look at an English Electric Lightning aircraft that the groundcrew had parked in a nearby hangar. This iconic single-seat aircraft incorporated two massive engines mounted one above the other. It looked as though the designer installed the cockpit on top of these enormous engines as an afterthought in typical British fashion.

When I tried to ease myself in, I was surprised by the cramped space and found it impossible to straighten my legs below the instrument panel to sit down. The ejector seat was just too far forward for me. Therefore, it was no surprise to hear that only pilots shorter than 5 foot 6 inches could fly this aircraft. However, it was still a spectacular and remarkable aircraft designed to intercept Russian bomber aircraft during the 'Cold War' period of the sixties. It could accelerate to Mach 2 (twice the speed of sound), in a vertical climb straight after take-off, and reach 40,000 feet from sea level, in three and a half minutes. But the limiting factor of this aircraft was its high operating cost and its limited endurance of only forty-five minutes.

At the end of the Sea Vixen course, I came away with glowing assessments on my qualification certificate, which I proudly pasted in my logbook. However, this also meant that circumstances were about to wrench Tish and me cruelly apart. After only four months of blissful married life, she would soon be flying back to her parent's home in Rhodesia.

I had received a posting to No 893 Squadron, which was operating off HMS *Victorious* in the Far East, and as the time drew near for us to part, I started to wobble again. I so wished I had chosen to join the RAF instead and not have to endure the extended periods of separation. Tish's pregnancy precluded any chance of her following me around the various ports, using the 'RAF indulgence passage' facility. Other married couples managed to take full advantage of this service, which allowed them to get together each time the ship sailed into ports in Singapore, Hong Kong, Sydney, or the US base of Subic Bay in the Philippines.

We spent Christmas and New Year in London. With hearts breaking, we shared the bitter-sweet time, window shopping in London's magical winter wonderland. As we walked along desperately clutching gloved hands, feathery snowflakes floated onto our cheeks. Already dark by 4 o'clock, the evenings were bitterly cold. Magnificent Christmas decorations lit up every street and shop window, and my heart ached as the separation loomed ever closer. I fluctuated between denial and despair, envying the other people so busily engrossed in their ordinary stable, happy lives and felt numbed by the unknown impending future that awaited us.

Over the last four months, Tish had become my strength and stability, and I wished I could hold on to the *status quo*. But of course, that was impossible in the Fleet Air Arm.

5 AIRCRAFT CARRIER OPERATIONS

HMS Victorious

Sea Vixen landing on HMS Victorious

The Navy booked my passage, and the time had come for Tish and me to part after four short months of married life. As I walked across the tarmac at RAF Lyneham in the cold night air, I wore my bulky sheepskin jacket, which would then have to sit in my tin trunk for the next eight months in the tropical climate of the Far East. Climbing on board the four-engine turboprop Britannia aircraft, affectionately known as 'The Whispering Giant', I looked back at the lonely sight of my pale and beautiful wife. She would soon be departing on board an RAF indulgence flight and stay with her parents in Rhodesia while I was at sea. She still looked as young and supple as a teenager, expecting our first child.

The Britannia was refuelled at the British-occupied port of Aden and then made a short stop at RAF Gan, which is one of the Maldives Islands in the Indian Ocean. It went on to land at RAF Changi in Singapore. Once there, a career gunnery officer, Lieutenant Alan Tarver, met me. He had previously volunteered to become a pilot, and we had finished our conversion onto the Sea Vixen at RNAS Yeovilton together. Now he drove me to the Singapore Docks, where I got my first look at HMS *Victorious*, which was tied up alongside the quay. As I walked up the gangplank, the massive grey bulk dwarfed the nearby dockside sheds and cranes. This was to be my home and workplace for the next eight months. The very next day, the ship left Singapore docks and headed for the US Naval base at Subic Bay in the Philippines.

Before my first Sea Vixen flight off HMS *Victorious*, I went on a familiarisation flight in the back seat of a Buccaneer low-level bomber of No 801 Squadron. These were the other jet aircraft operating from HMS *Victorious* at that time. I sat in the observer's seat, which the designers had raised above the pilot's seat to allow an unobstructed view over his head. Before becoming airborne via the steam catapult launch, the deck crew pulled the Buccaneer into high nose attitude with the nosewheel off the deck. The raised in-flight refuelling nozzle made the aircraft look like some giant stinging insect, with its sharp nose pointing skywards at the precise take-off angle. When ready for launch, the pilot withdrew his hands from the control column and throttles and laid his arms on the provided rests. The aircraft's posture was at the correct angle for take-off until it had accelerated to a safe flying speed. This procedure avoided the pilot's risk of inadvertently pulling back on the throttles or controls during the 4G acceleration. It was still surprising to feel the aircraft sink below deck level before the speed built up enough to climb away, even at full throttle. After that flight, I was deemed ready for my first Sea Vixen flight off the carrier.

News spread through the ship that a rookie pilot was about to go on his first launch and recovery. This event brought onlookers rushing to the *'Goofers Tower,'* which was a favourite vantage point, overlooking the flight deck. When I was on the flying programme for the next few days, this balcony was jam-packed with onlookers that looked like a committee of vultures, waiting in morbid anticipation of some gruesome entertainment.

Meanwhile, when I was not on the flying program, I took every opportunity to watch the on-deck operations from the same place.

Thankfully, my first flight went off without a hitch, and on the landings over the next few months, I managed to catch the favoured second, or third of the four arrestor wires. Moreover, on most occasions, the nose wheel was on or reasonably close to the centre line. The catapult launch itself was the highlight of each flight. It was an exciting way to start a sortie, and such an exhilarating experience, that the Navy pilots referred to it as, "The next best thing to sex!" Getting into the air also allowed me to escape the constant nauseating movement of the carrier. As the aircraft climbed away, I would breathe a sigh of relief as my world stabilised when we cut smoothly through the air. I now felt at home and back in my element.

When I had been on board for just over two weeks, I spent a pleasant evening chatting to one of No 893 Squadron's senior pilots, at the wardroom bar. I discovered during the conversation that he had over 3,000 hours flying and was also a qualified AWI (Air Weapons Instructor). He was a humble man called Bill, who reminisced about his Norwegian wife, who went to stay with her parents in Oslo every time he was at sea.

At that time, drinks were a penny a tot, but no cash ever crossed the bar. Each officer was allocated a bar number and only paid for his own drink. Instead of paying for a round, one would ask, "Can I write a drink for anyone?" and then give each person's number to the barman.

As usual that evening, I did not stay in the wardroom long before heading back to my confined cabin space. Bill and I flew a formation sortie in two Sea Vixens the following day. But for some reason, the Squadron technicians had not fitted my aircraft with the under-wing fuel tanks. With less than the usual fuel in my plane, I had to recover back on the carrier with the Buccaneers, as their sorties were limited to only 50 minutes.

When I was safely on board and busy climbing out of the cockpit, I fully appreciated how turbulent the conditions were. With only two weeks' experience under my belt, the weather had drastically deteriorated by the time I returned to land on the deck. The carrier was now right on the limit for landings in rough seas, with the deck pitching up and down thirteen degrees. The turbulent waves resulted in the ship's back end rising and falling by twenty feet (six metres). On the landing approach, apart from keeping the aircraft on the extended centreline, I ignored the deck while focussing all my attention on the gyro-stabilised landing sight. The plumes of spray that shot over the ship's bow drenched the flight deck, and deckhands were in danger of losing their footing in the whipping froth. The vessel continued to rear up and fall headlong into the swell.

After signing off, I made my way to the *Goofers Tower*.

Staggering along, I was thrown about and clutched at any available handhold. When I arrived on the balcony, I was just in time to see Bill's aircraft coming in on the final approach. I was immediately concerned when I noticed significant pitch changes to his aircraft's nose attitude. The continual rise and fall of the plane's attitude could only mean he was struggling to maintain a steady approach speed. Simultaneously, the aircraft's wings banked sharply from side to side, which indicated that Bill was overcorrecting as he fought to keep lined up on the centre line. I could empathise with his battle against the gusting conditions. Then a few hundred feet off the deck, he called off the approach and climbed away for another circuit. One of the nearby spectators mentioned that it was the fourth time Bill had aborted his approach to land on short finals.

On this fifth attempt, I felt the tension building up. Once more, word had spread through the ship that a pilot was experiencing difficulties, and in no time, spectators took up every available space on the *Goofers Tower.*

I became exceedingly anxious when I saw Bill struggling with his lineup and speed control again. Holding my breath, I willed a safe landing as the aircraft roared in towards the deck. Then surprisingly, just a few feet off, he pulled the nose up for another go-around. I now worried that his dangling hook might snag an arrester wire, pluck the aircraft out of the air, and slam it onto the deck with a catastrophic consequence. But as these thoughts raced through my mind, the plane passed level to the *Goofers'* platform. I gasped when I saw the right wing suddenly drop inexplicably to a ninety-degree angle. I prayed that Bill would be able to check the aircraft's roll in the inverted position, push forward on his control column and climb safely away. Frozen to the spot, while everything seemed to happen in slow motion, I watched in absolute horror as half the aircraft's wing ripped off when it clipped the folded wing of a Sea Vixen, tucked tightly between other aircraft on the catapult launch area.

Realising all hope was now lost, I turned away in shock and covered my face with my hands. I wondered if the crew could save themselves by ejecting out sideways from the aircraft. But when I looked back, I saw debris scattered over a large area of sea on the starboard side, and saw the observer's blue 'bone dome,' sweeping past the side of the ship. It was apparent that there would be no chance of finding any survivors.

Witnessing this accident had a profound effect on me, as it drove home just how dangerous it was to operate off an aircraft carrier, especially considering Bill's flying experience. However, the Board of Enquiry discovered that the night before, Bill had consumed thirteen brandies and ginger ale (known in the RN as *'horse's necks'*).

Studying the film taken to record every launch and landing, it showed by the continual movement of the control surface, that Bill had attempted to fly his plane right up to the point it crashed into the sea.

To be honest, I did not take well to life onboard the carrier. I spent a large portion of my time in the ship's bowels, with no reference to the natural horizon. The constant rise and fall of the vessel made me feel perpetually seasick. Moving from one part of the ship to another during periods of rough seas, intensified my nausea. Under these conditions, I would stumble along from side to side, as the ocean's enormous power rock and rolled the carrier. One minute I would struggle in an uphill climb, then the next moment, gravity would pull me down in a headlong rush.

It also took me weeks to find the shortest route from one part of the ship to another. Unlike *Edinburgh Castle,* in which I had recently travelled from Cape Town to Southampton, HMS *Victorious* had no passageways running the ship's length. One had to climb and descend dangerously steep narrow metal stairways in stifling heat, and zigzag from one direction to another, while ducking through small metal hatchways. The luxury of sweet-smelling air conditioning was also missing, and there was no wood panelling to conceal the noisy acrid smelling pumps and the steam hissing from whitewashed cladded pipes.

When not on the flying programme, kitted out in the sensible tropical uniform of white shorts, an open-necked shirt, and brown sandals, I managed to escape to a small, recessed area on the side of the ship. This spot exposed every outside surface to the elements, and the rails were always clammy with sea salt. Standing out there, when the wind was not blowing choking black smoke that often billowed from the ship's funnels, I could breathe in gulps of fresh air. There, I could lose myself as I watched seagulls circling, swooping, and plunging into the endless grey seas that slid past five stories below. The scene was often so dreary that I yearned for the sight of land and trees. On the odd occasion, the chance sight of palm trees on an isolated tropical island did much to lift my spirits.

Walking around the carrier, I often passed one of the ship's complements of two thousand, as they went about their duties. As we squeezed past each other, along the labyrinth of walkways, hatches, and ladders, there was never a hint of acknowledgement or greeting, as the Navy had wisely dispensed with saluting when at sea.

During launch and recovery, the ship came alive as it vibrated and charged headlong into the battering wind. The efficient buzz of noise and activity on the flight deck was enthralling. The blast of the jet engines was deafening as aircraft crisply pirouetted around the deck, wings neatly folded up and looking like some grotesque butterflies in a crazy barn dance.

The scene during the aircraft launch and recovery was exceptionally breathtaking. Marshallers synchronised aircraft movement to the launch pad by communicating through radio headsets. They would impatiently stamp their feet to direct the plane in and out of tight spaces on the deck. The pilots would respond to the brisk hand signals with throttle bursts, while urgently winding the nose wheel steering and abruptly jabbing the brake pedals. On deck, marshallers had absolute authority, which absolved the pilots from any responsibility to avoid obstructions.

The marshallers guided the aircraft onto one of the two steam catapult launch pads, where the pilots briskly taxied up against the raised hydraulically operated chocks. Once in position, the thick-cabled metal bridle was swiftly looped round the steam-operated shuttle and attached to horns below each wing. The holdback bar was set at the precise breaking strain and simultaneously raised from the deck to the aircraft attachment point. The deck controller then retracted the chocks and raised the blast deflector.

In obedience to the launch controller's hand signals, the pilot lowered and locked the wings to the extended position, set the flaps, and opened the throttles to take-off thrust. When everything was ready for the launch, the pilot would give a sharp nod before pushing his head hard back against his headrest. The launch controller then dropped into his distinctive pose, stretched on one knee, and pointed his whole arm in the take-off direction. After a moment's delay, the pilot would be pushed back into his seat, and with a whooshing blast of steam, the catapult would accelerate the aircraft from a standstill to flying speed in just sixty metres.

The Sea Vixen's flight crew included the pilot, and an observer behind a console and radar screen. To the right and below the pilot was the observer's cockpit, which he entered from above through a solid hatch that closed flush with the top of the fuselage, while the pilot's protruding windscreen and canopy overlooked the left side of the aircraft nose.

The observer was primarily responsible for all aspects of navigation. But, whenever we did low-level navigation over land, and since the observer could not see the ground ahead, I chose to do the navigation myself, which I had done in single-seat aircraft before. When planning this sort of exercise before launch, I would first draw the track onto my map, and then use it for the map-reading. But it was another matter when carrying out high-level navigation, as the observer could then use available ground beacons, and pick up the shape of coastlines on his radar.

When carrying out a landing approach on the Sea Vixen, the pilot could select an audible tone on his headset. This tone changed with the increase or decrease of the aircraft's speed, to help him hold the correct airspeed for the aircraft's weight. This audio was an innovative system that measured the wing's angle of attack and gave a steady note at the right approach speed for the aircraft's weight. When the airspeed was too high, the pilot would hear a quick pulsating high-pitched sound. But when the airspeed dropped below the correct airspeed, the pilot would get a slower and lower-pitch sound.

However, although I could see this benefit in a single-seat aircraft, I preferred to have the observer call out the airspeed in this two-crew situation. I found it easier to hear how I was doing from the pitch and urgency in his voice!

The observer also played an essential and comforting role during steep rocket attacks over water. It was a great advantage to have him call the height above the water, which allowed the pilot to devote all his attention to the gunsight and keep the correct attack speed. The observer usually made these callouts when delivering weapons on a 'splash target,' which was dragged 1,000 metres behind the ship.

My first observer on *HMS Victorious* was Lt. Dick

Gravestock, who was No 893 Squadron's Senior Observer. He told me about the accident he had previously experienced in a Sea Vixen, which had crashed into the sea off the carrier's front. This accident followed a 'cold shot' during the catapult launch when the steam-powered shuttle failed to do the job of accelerating the aircraft to its flying speed. With inadequate speed for flight, the plane went off the front at taxi speed and plunged headlong into the sea. When Dick had jettisoned his overhead hatch, seawater gushed into the observer's cockpit with such force, that the water ripped off his oxygen mask, and he found himself trapped in his ejector seat. His pigskin flying gloves had quickly absorbed moisture, and this made them too slippery to turn the seat belt's 'quick release box'. Luckily, a diver from the rescue helicopter was able to help him out of his seat harness and leg restrainers and pull him out of the aircraft before it sank underwater. He never actually mentioned what happened to the pilot, who I presumed managed to get out safely.

Walking out to the flight deck for our first flight, Dick calmly informed me that I would be on my own if the speed dropped below 115 knots on the landing approach. Thankfully, I never gave him any reason to eject on that flight, and we continued to fly together for over a month.

After Dick, Mike Parker became my permanent observer. He was a stocky young man with abounding self-confidence as a ladies' man. To me, he looked more like Pinocchio or some other cartoon character, but he seemed to have raw animal sexuality, which appealed to the opposite sex. However, what I enjoyed most, was his enthusiasm during aerial combat. Without visual reference to the aircraft's attitude to the natural horizon, the moment I began to haul the plane into a high 'G' manoeuvre during a 'dog fight', he would start rattling off the airspeed. I found this quite remarkable when I considered that the observer, sitting in the claustrophobic enclosed cockpit, could not see the outside world. I can only assume he derived vicarious pleasure from my continually unbeaten success.

One day on a navigational flight, while he was busy plotting our position, and his attention was focused on his map, I set the aircraft into a gentle descent from 28,000 feet, down to sea level. When the plane was skimming across the chopping sea, with tossing white horses almost touching our belly, I said, "Hey Mike, look out of your window!" Mike saw the waves almost lapping against the plane's underbelly, and I was childishly amused by his panic-stricken scream when he glanced out of the small window below his right shoulder.

A gyro-stabilised optical landing system on the deck touchdown area's left-hand side, helped the pilots maintain the correct glide-path angle when approaching the aircraft carrier. The former system they had used was called the 'meatball,' as it utilized a ball of light reflected from a concave mirror. But by the time I was flying, an updated system had replaced it. This system incorporated a vertical row of light bulbs and a datum crossbar of green lights. When on the correct glide path, the pilot saw a white light in line with the green lights' datum bar. The white light went up and down if the aircraft went above or below the correct glide path. When the approaching plane sank below the proper approach angle, the white light would be seen below the datum bar and turn red when the plane sunk dangerously below a safe approach angle.

As part of the landing checks, the pilot would lower the underbelly arrestor hook when the undercarriage was down and locked. Hopefully, the arrestor hook would catch one of the four hydraulically dampened cables when the aircraft finally thundered onto the deck. These cables were held off the deck by compressible leaf springs and would rapidly decelerate the aircraft to a standstill in about one hundred metres.

The 'wires' that crossed the deck were strong,

hydraulically dampened, cables. The number three wire was the target cable, which the aircraft caught if it remained on the correct glide path angle right down to the deck. There was then a wider gap between the target and the fourth emergency wire.

A camera on the captain's bridge, recorded every aircraft landing and catapult launch. The safety officer would later study the films to analyse accidents or any dangerous tendencies. The Chief Pilot would then inform pilots of any developing bad habits and discouraged them from catching the first wire by lunging forward as they crossed the 'round down', which practice, the pilots would call *"getaboarditis.* Over the years, many aircraft had smashed in short. Amazingly, the slightly rounded area at the beginning of HMS Victorious' landing deck, bore witness to this practice by several visible dents on its thick steel plating. Pilots adopted this practice to avoid missing all four wires and doing a 'bolter'. In this case, they would need to select full throttle, get airborne again and re-join the circuit for another attempt.

Living on board *HMS Victorious*, I was continually homesick and yearned to be with my beautiful wife. Between flying sorties and mealtimes, my loneliness led me to withdraw from the other pilots and spend long hours, just lying on my cabin bunk, situated way below the waterline and somewhere in the middle of this lumbering hulk. I would lie there for hours, with the Beatles music blaring out, "She loves you, yea, yea, yea" while I stared at the brightly coloured photographs of the Rhodesian bush that I had cut out of a calendar and stuck to the metal bulkhead.

One entered the cabin through a sliding door. Once inside, half the available space was taken up by the raised bunk bed, above a set of drawers and a fold-down desk. When the desk was pulled down, there was just enough space to squeeze in, with the back of the chair touching the opposite bulkhead.

I relished the times that the ship sailed into port for a couple of weeks and the squadron would deploy aircraft to one RAF base such as Kai Tak in Hong Kong or Changi in Singapore. Whenever we stayed at one of these officers' messes, it gave me the chance to relive my previous air force life, so recently thrown away. However, it was a bittersweet experience, which made me rue that I did not join the RAF instead of the Fleet Air Arm. The RAF was more familiar and painfully reminiscent of my life at Rhodesia's Thornhill Air Force base. At these RAF bases, life seemed ordinary, especially when I visited the camp swimming pool and could hear the children's screams and happy laughter. It was music to my ears when I heard them frolicking in the water, under blazing sunshine.

At that time, Singapore was still part of Malaya, and when the carrier docked there, while the other pilots returned to their shipboard cabins, I could not wait to escape my claustrophobic space and book into the Air Force Mess for a few days' break.

The road outside Changi was a thriving roadside market with stalls that sold everything at duty-free prices. Here, most of us took the opportunity to buy fancy Sony and Akai reel-to-reel tape recorders and speakers. The vendors were also happy to copy any LP ($33\frac{1}{3}$ rpm long-playing record), onto a tape, for a reasonable price. I bought a compact Sony tape recorder, which was carried like a small suitcase with two detachable speakers.

When in Singapore docks, I sometimes took the opportunity to go into the city on what the Navy called a "run ashore". On these trips, I managed to see the original Raffles Hotel and Boogey Street, made famous by men dolled up to look like beautiful women who would tempt unsuspecting tourists. However, I found Singapore extremely hot and muggy, especially in my 'drip-dry' shirts that clung uncomfortably to my sweat-stained torso. Since then, I have only bought cotton shirts. Travelling the city's outskirts, we often passed overcrowded shantytowns, crowded, rusty corrugated iron shacks and an abundance of wooden plank bridges that crossed areas of putrid-looking water. But what always amazed me was to see immaculately dressed Malay girls emerge in impeccable outfits and shiny high-heeled shoes. I thought they must surely be some of the most beautiful race of women on earth with such delicate features, almond-shaped brown eyes, and neat figures.

On one occasion in Singapore, we were sent on a survival exercise where we spent a night in a steaming tropical jungle. Here we learnt how to build raised beds to escape all manner of reptiles and crawling insects. On the second day, we did a short navigational exercise. Here it was necessary to continually take compass bearings to keep in a straight line as we hacked our way through the impenetrable jungle and waded through leech-infested rivers. Each time we emerged, we found leeches clinging to our flesh, and the only way to detach them was with the tip of a glowing cigarette. Despite all this, I found it strangely refreshing to experience the free flow of sweat in the hot and humid conditions, while being shielded from the direct sunlight by the tree canopies that stretched hundreds of feet above us.

One evening in Singapore, I went for a meal in town with a young Squadron observer. Returning to the docks by taxi, he asked if we could stop at a brothel. I was somewhat surprised by this request and said it was up to him, but to leave me out of it. He then asked the taxi driver if he knew of someplace. On arrival at a seedy-looking establishment on the city outskirts, we were ushered into a waiting area and presented to a sullen group of women, who were still in their dressing gowns with unkempt hair and dragged from their sleep. To my mind, they were undoubtedly the most uninviting bunch of women I had ever seen, but my young friend was pleased to select one of them. He followed her to a screened-off area without soundproofing. As the other girls and manager viewed me with dull suspicion, I felt embarrassed and uncomfortable while waiting for him. I could not get out of the place soon enough, but as we continued to the docks, my companion was now feeling self-conscious and swore me to secrecy. Sadly, a little while after this episode, when I was on compassionate leave in Rhodesia, he was killed when the Sea Vixen he was in, lost an engine straight after launch, and he could not eject from the aircraft before it ditched into the sea.

The duty-free port of Hong Kong was also a place where we could find real bargains. Whenever the carrier dropped anchor in the harbour, Chinese junks would come alongside to collect scraps of food and other refuse. While there, I found a tailor who offered to make me a dinner jacket and an embroidered golden silk cheongsam for Tish for the equivalent of twenty pounds. He took my measurements, and two days later, the dinner jacket fitted like a glove. Later, when Tish wore the figure-hugging cheongsam, she looked stunning.

My colleagues and I often visited Kowloon's mainland to enjoy a sumptuous meal of Chinese fried rice with *egg fu Yong,* spring onions, and other unknown greens. This cost one Hong Kong dollar, which was then the equivalent to one shilling and three pence. We often visited a fancy floating restaurant alongside the boat people, who lived on one of the jammed-together, countless, small vessels. At other times we bought tasty meals cooked over glowing pavement braziers while we waited.

One day, I went by myself on a ferry from Hong Kong to Kowloon, where I walked miles away from the quay up a long straight street called Nathan Road. After a while, I felt conspicuous as I was the only Caucasian in the vicinity and towered over the thronging locals. When I felt hungry later and entered a restaurant with fish swimming in a display tank. But, as no one spoke English, I pointed out an interesting-looking pink fish but was disappointed when all they did was boil it and present it on my plate. It was utterly tasteless and unpalatable, and I could not finish it.

I then found a barbershop. But once again, no one could speak English. However, undeterred, I sat down for a haircut. But instead, the barber subjected me to real Chinese torture. When he applied the shampoo to my scalp, he massaged it in with long fingernails that brought tears to my eyes. This process continued for an unbearably long time, but I was far too shy and embarrassed to protest.

On another occasion, a group of us took a hydrofoil to Macau's Portuguese colony, which was even then, a gambler's Mecca. The trip across was exhilarating. I found the architecture surprisingly like the Portuguese port of Beira in Mozambique, where I had spent many long weekends and holidays soaking up the sun and enjoying the warm waves of the Indian Ocean. We did a lot of sightseeing, but I cannot recall participating in any gambling activities. We spent the night in a cheap hotel, where washing was draped over every balcony of the surrounding apartments, resounding with the irritating strains of Chinese music.

After a few months in the Far East, we met with our sister carrier, HMS *Ark Royal*, for a combined naval exercise. The Sea Vixen squadron that operated off this carrier, flew the Mk II, an updated version of our Mk 1. The two versions' difference was immediately recognisable by protrusions forward of the leading edge of each wing. These ugly-looking extensions, which stretched ahead of where each tail boom joined the aircraft's wings, carried extra fuel tanks. But although this increased flying range without the need for drop tanks, they also restricted the pilot's left and right peripheral vision.

I came across a lone Sea Vixen off Ark Royal during one airborne exercise. We operated on different radio frequencies on unrelated tasks, but we went into a spontaneous dogfight when we saw each other. He lost sight of me in no time, and I was on his tail. Unaware of this, he continued maximum G turns while craning his neck around to find me. Unseen, I moved into close formation on his starboard wing and saw his head continue to swivel as he strived to find me.

Suddenly, I got the fright of my life when he unexpectedly reversed the turn and rolled his aircraft towards me. As I instinctively pulled up and away from him, he flashed underneath me, narrowly missing a mid-air collision. I was obviously in his blind spot, and he was oblivious to his close encounter with death. He then flew off to rejoin his mother ship, and I never disclosed our unauthorised escapade and how close we came to an inexplicable tragedy.

As a policy, the Navy discouraged doing 'greasers' on land-based runways, a term used among pilots to describe soft landings. But after flying out of Kai Tak Airport for a couple of weeks, I learned the wisdom of this policy the hard way. I enjoyed flying from a proper runway again and began to hold off to slide the aircraft smoothly onto the runway. Until then, when doing carrier landings, I had never experienced a 'bolter' by missing all the arrester wires. But, when the time came to re-embark, I failed to catch any of the four wires, despite thinking I had maintained a steady nose attitude right onto the deck. I was taken by surprise when the immediate and satisfying decelerating pressure on my shoulder harnesses failed to occur.

Instead, the aircraft continued at breakneck speed across the deck, and just as I started to close the throttles, I was shocked to realise what had happened and had to ram the throttles to full power. Fortunately, the Sea Vixen had excess engine power. My plane leapt back into the air at flying speed, and I was able to climb away for another circuit. Shaken by this unexpected occurrence, it took me five more attempts before I forced myself to push forward on the control column as I crossed the round down. But by then, my fuel was down to the reserve needed to divert back to Hong Kong. Once again, the story that a pilot was having 'trouble' sped through the ship. By the time I caught a wire with my confidence in tatters, the *'Goofers Tower'* had thronged with onlookers, relishing the unexpected entertainment. Climbing down to the deck, I felt sheepish but relieved that I had avoided the further humiliation of diverting back to Hong Kong.

Whenever the carrier steamed into port, the ship put on a neat display of sailors evenly spaced around the flight deck. The ship's crew looked pristine in crisp white uniforms, with the aircraft parked on display.

The captain's staff always laid on a formal cocktail to entertain the local dignitaries and other guests to conform with tradition. This function took place in the aircraft hangar below the flight deck. The ship's officers would line up along one side of the hangar to welcome guests as they arrived. They presented a stunning picture, dressed in their 'Red Sea rig', comprising starched white longs, white open-necked short-sleeved shirts with epaulettes, white shoes, and colourful cummerbunds, denoting their squadron or section. The ship's band added a festive atmosphere when it made a dramatic entrance, while descending on the giant aircraft lift from the flight deck. The hosting officers plied the welcomed guests with drinks by using their personal bar numbers. This arrangement seemed extravagant, but it was cheap at only a penny a tot. As officers chose which guests they wished to entertain, some were inevitably skilled at making the right choices. Their prime objective was to profit from the anticipated reciprocal hospitality during their time in port. These officers earned themselves the uncomplimentary title of "Baron Stranglers".

On one of these occasions, when the ship docked in Hong Kong, the actress Ursula Andress was on a film location and was somehow enticed to visit the aircraft carrier. She was already a sex symbol following her famous and iconic appearance in 'Dr No', the James Bond movie. In one remembered scene, she displayed her well-proportioned figure in a scanty white bikini as she stepped from the waves onto a Caribbean beach.

When she came on board HMS *Victorious*, she was ushered straight to the open-sided quarter-deck, situated at the vessel's stern. The officers swarmed around this beautiful young film star unashamedly to admire her generously displayed cleavage. Although undeniably beautiful and shapely, what surprised me was how petite she was. From the film, I had somehow pictured her being tall and willowy. But after a short while, she asked to meet the sailors, as I imagine she was bored with the toffee-nosed officers and wanted to bestow her fantastic assets upon rawer, down-to-earth, and lusty admirers.

Understandably, the officers were reluctant to let her go, but surprisingly she invited my observer, Mike Parker, to join her that evening. I never asked how it went, but I heard that when he returned in the early hours of the morning, just before the welcoming party piped the captain aboard, and in full view of the boarding party, he hung puking over the ship's railing.

One of the first ports of call after I joined *HMS Victorious*, was to the US Naval Base at Subic Bay, in the Philippines. At the welcoming cocktail party, my observer and I had the good fortune to host a pleasant American couple. The husband was a teacher at the military base school, and his wife was a friendly, down-to-earth person who proudly told us she was a quarter Red Indian. After the cocktail party, they invited us to join them for dinner in Olongapo, a sleazy local village that reputedly had more bars and nightclubs in one mile than anywhere else in the world. However, the captain's office advised the ship's crew not to venture beyond the main street. After being invited ashore for dinner, our hosts took us to one of the many nightclubs, where I was shocked to see strip tease acts, performed by girls as young as eleven. The following day we heard that someone had stabbed a sailor when he roamed too far.

Four months later, we had just gone into port at Singapore docks when I received a telegram from my father-in-law informing me that our baby had died at birth. Tish was reportedly okay, but the Navy immediately granted me compassionate leave. The admin office booked me on an RAF 'indulgence passage' to Rhodesia via Bombay, where I spent one night at a hotel. Coming from Rhodesia, where the white population owned similar-looking homes, I was surprised to learn from the taxi driver that Indians owned all the neat suburban houses that we passed. He then pulled a fast one on me by claiming he had no change but promised to square up when he came to fetch me the following day. Still naive, I took his word for it. But of course, he failed to show up, and after a long wait, it eventually dawned on me that he had no intention of coming back, and I was obliged to call for another taxi to take me to the airport.

At Salisbury Airport, Tish's father met me and took me straight to the hospital in Gatooma. Tish was in a private ward, and my heart broke when I saw her looking so pale, beautiful, and shell-shocked after her ordeal. I just held her close as we wept in each other's arms. I grieved with her and for her. At the same time, my heart so yearned to return to this beloved country of my birth that it cemented my decision to resign from the Fleet Air Arm.

On my return to the ship, I immediately submitted my resignation. I was relieved after taking this positive step to overcome the cause of my suffocating depression.

It also gave me some respite and peace of mind to think there was light at the end of this dark tunnel. I was then able to relax in the knowledge that I just needed to complete this tour of duty before Tish, and I could pick up our lives together.

I started participating in sports and other activities with the squadron aircrews with a more positive attitude, and became friends with Don Mckenzie, one of our Sea Vixen observers. Sporting a curly blond beard, he had a real zest for life and a broad flashing smile, which displayed a row of perfect and evenly spaced white teeth. He had an effervescent personality and was everyone's friend, and he kindly invited me to join a syndicate to purchase a little speed boat. Then, whenever the opportunity arose and the ship laid anchor in the South China Sea, our group would launch the boat from a gangway and take a picnic basket to one of the small deserted tropical islands. There we would spend glorious days swimming, sunbathing, and water skiing. Without Don, I would probably have spent more time just lying on my bunk.

I also became a member of the Squadron's volleyball team which we played in the partially lowered aircraft lift well. My extra height was an added advantage, and it allowed me to get some exercise in the sun and fresh air. However, this was occasionally spoilt when the stokers discharged billowing, thick black smoke from the ship's funnels. This smoke would sometimes spew choking carbon particles over the flight deck in still wind conditions.

With only artificial light in the carrier's bowels, there was no way of knowing the external weather conditions or the time of day. Then one day, I overslept by a few hours and staggered with a towel draped around my waist to the 'heads' (ablution block). Thinking it was the middle of the night, I was surprised to see other officers walking about fully dressed. In a panic, I quickly donned my flying overalls and rushed to the briefing room, to see if the flying program had scheduled me for an early launch, but to my relief, I found that I was on a later flight.

On the carrier, cabins doubled as officers' workplaces. The size of the accommodation depended on the officer's rank. Although I had no reason to visit the captain's quarters, I imagine he had more than enough space with a living area and a window or two included.

Following the well-known historical event of the Mutiny on the Bounty, the Royal Navy decided to appoint every officer as a 'Divisional Officer' when onboard an RN ship.

This function made him responsible for all non-commissioned officers, or seamen welfare matters, in his division. This practice allowed the lower ranks to vent any frustrations and opened channels of communication between the different levels, which has since proved to be an effective way to nip any festering discontent in the bud.

I had a few ratings in my division and enjoyed the opportunity to relate with them. Occasionally, one of them made an appointment to discuss domestic or other problems. The ratings in my division usually learnt about issues from the letters they received from home. The family could keep in touch by the regular mail runs delivered by a Gannet aircraft. Each officer held a personal file for every member in his division, which he would then keep up to date. Every issue brought to the divisional officer was discussed and redirected to one of the ship's administrative officers. They would generally be either the ship's doctor or one of its pastors. Depending on the specific problem, the section would arrange for a qualified officer to rectify any problematic situation in the UK.

When I saw Tish in Rhodesia, I mentioned how much her letters had meant to me, and after I got back to the carrier, she wrote to me every day for our last four months of separation. Every few days when the mail arrived, to everybody's amazement, I always received a handful of letters.

The Squadron assigned two more duties to the junior aircrew officers. The first was to oversee the daily rum issue. At this time, catering staff watered down full-strength Jamaica rum into brass containers. The strong, sweet-smelling liquid would be carefully measured into different-sized brass ladles and scooped into brass churns for each mess. The higher-ranked NCOs received less diluted tots, but there was always some excess, which the staff measuring the tots would toss out across the deck into a drain hole. Although I never drank 'hard tack', the thick sweet-smelling rum would compel me to go directly from there to the wardroom bar and order a glass of rum and coke!

The other duty was to watch when the Squadron maintenance technicians towed an aircraft between the flight deck and hangar to carry out any work. This procedure ensured more care was taken, and in the event of a mishap, it minimised any opportunity to pass the buck.

When I was on *HMS Victorious*, a couple of technical tragedies occurred. The first was when a deck operator was sucked into a Sea Vixen's intake when he had walked too close to the engine while engine technicians opened the engine to full takeoff thrust during a routine maintenance check. The suction was so powerful that the rating was pulled off his feet and ingested headfirst down the engine intake. Before there was time to shut down the engine, the victim's lungs had collapsed, and the impeller blades had trapped his fingers. The only way to free the corpse was to send the thinnest Squadron-tech down the intake. He had to squeeze past the lifeless body and cut his fingers loose. I later heard that the traumatic experience had caused this youngster to suffer a nervous breakdown. The other tragedy involved a Sea Vixen in the aircraft maintenance hangar.

It happened while two armourers were inspecting the pilot's ejector seat. One armourer stood on the access ladder against the fuselage to lean over the pilot's seat, while the other stood on the ejector seat itself. Part of the inspection was to remove the ejector seat's safety pin, and when he removed it, some defect caused the firing pin to strike the cartridge with a loud "Click". This sound alerted one of the technicians, that the rocket was about to fire the ejector seat out of the cockpit.

Standing on the seat cushion itself, the armourer, who had immediately realised what was about to happen in just over one second, did a backward somersault to the hangar floor, over ten feet (3.28 meters) below. Unfortunately, the technician leaning into the cockpit, did not react quickly enough, and the seat struck and killed him instantly as it shot up the rails and smashed into the overhead bulkhead. Strangely, the surviving technician, who only suffered a few bruises, had decided the previous day what he would do in such an eventuality.

On my introduction to night flying off the carrier, I experienced two scary situations. The first occurred late one afternoon when the Squadron had me on the program to carry out my first landing during twilight. Flying at this time would allow me the chance to familiarise myself with the deck lights before it became dark. However, the radar controller looked at the wrong aircraft on his screen when he gave me turns to vector me for a CCA (carrier-controlled approach).

It was already pitch-black by the time he eventually directed me onto my final approach. To my trepidation, the intended twilight landing had become the real thing. The introduction to landing on the deck at night was an unexpected and alarming experience. The ship looked like a matchbox, suspended in black nothingness. While I was descending on instruments, everything seemed to move in slow motion as it took forever to get closer to the miniature red centreline lights. Then it suddenly felt as I imagined it would be if I were sitting in the front of a train rushing into a tunnel at breath-taking speed. I was very relieved to feel the comforting strain against my shoulder straps as the hook caught one of the arrester wires.

The second incident was on my first catapult launch at night. After completing my pre-take-off checks, I was supposed to select the navigation lights to the 'steady' position. Instead, I put them straight to 'flashing,' which I should only have chosen when I was "ready for take-off" and had set the throttles to the full-power position. The moment I realised my mistake, I froze with shock and considered slamming the throttles to full thrust and hope for the best, but realised the engines could never spool up fast enough. The thought crossed my mind that I might need to eject if the controller dropped his wand to signal, "Clear for launch". In which case, although the shuttle would still operate correctly, there would be insufficient power to accelerate the aircraft to its flying speed, and it would merely plunge off the front of the ship into the pitch-black sea, from a few stories up,

I switched all the navigation lights off in my state of panic and indecision and held my breath. Fortunately for me, and notwithstanding the thundering noise all around, the launch controller, wearing his radio headset, stayed in the loop, and continued to circle the green wand above his head. So, taking a deep breath, I sheepishly opened the throttles to the 'take-off' position and set the navigation lights to the flashing position when ready. Only then did the marshaller drop his wand, to signal for the launch.

Despite the shaky beginning, I soon became confident with night flying operations. Of course, this was an essential requirement since the RN had chosen the Sea Vixen as an 'all-weather night fighter'.

Over the months, I realised that the Fleet Air Arm had an unusual tolerance towards pilots with an exhibitionistic flare. This acceptance gave me a licence to vent my frustrations. I started to give an impromptu low-level aerobatic display at the end of each sortie. While the next launch took place, I performed this exhibition in the recovering aircraft holding area, which was off to the carrier's left.

On one occasion, when the launch was complete and before the other aircraft flew overhead to join the circuit, I did a high-speed low-level flypast across the ship's bow. The only reaction I got was from the Squadron's Chief Pilot, Lt Cmdr Bob King, who said with some amusement, "You gave the captain such a fright he spilt his coffee!"

I continued to excel in aerial combat, and the Squadron also gave me the Squadron's Reconnaissance Pilot role. They tasked me to take aerial photographs of ships and land-based infrastructure for Navy intelligence in this role. I managed to work out a system to take film from the sideways-oblique camera, which the designers had installed in a specially designed, left drop-tank. There were no instructions on the line-up of this camera, but after some trial and error, I drew an elongated box outline on the canopy's left-hand side, which did the trick. I enjoyed this assigned task, and I even had an opportunity to film a formation of Sea Vixens as it did a flypast over HMS *Victorious*.

Towards the end of my tour, we had just completed a combined naval exercise with a fleet of various American warships, including the first nuclear-powered aircraft carrier, USS *Enterprise*. I even had a chance to carry out a 'touch and go' on it, and I was impressed by the deck's sheer size. But what intrigued me most was to notice that the shipbuilders had covered the entire landing deck with wood. I was also amazed at how quickly they taxied each aircraft onto the aircraft lift and took it to the hangar below, straight after it had landed, to clear the deck before the next plane could land.

After this combined exercise, each captain chose an area to anchor their vessel in a large bay at the Gulf of Thailand's top end. Once there, Andrew Dalrymple-Smith suggested 'a run ashore' on one of the ferryboats. He was an eccentric but intelligent pilot who had been on the same Hunter course as me at RNS Brawdy in South Wales. On this Hunter course in Wales, the other students gave him the unflattering nickname "Harpic." This name came from a current TV advert for Harpic toilet cleaner as his course mates thought he was "completely round the bend". Our career paths had brought us together again, as he was now one of the ship's Gannet pilots, and I sometimes enjoyed chatting with him in the wardroom bar.

Once ashore, we caught a bus into Bangkok, which seemed to take forever. On the way, we passed endless Buddhist shrines and water buffaloes ploughing rice paddy fields. In the city, we spent some time sightseeing in the heavy humid conditions, and as I had brought my swimming costume and towel, I took the opportunity to cool off at one of the tourist hotels. We then decided to find somewhere to escape the oppressive heat and have a bite to eat. We passed a cosy little air-conditioned bar and restaurant, and when we heard modern hit songs on a jukebox, we decided to go in. Inside, we saw a flock of petite Thai ladies wearing short, figure-hugging cheongsams and had barely settled down to a tasty meal of rice and prawns when the place was descended upon by a horde of loud drunk, and brawling US sailors who were out for a good time. In my innocence, it took me a while before I noticed that sailors would periodically disappear with one of the sweet-looking doll like girls who would then return five or ten minutes later. As though nothing had taken place, each girl would continue with her meal of rice while she waited for the next sailor to invite her onto the small, darkened dance floor. After a while, some of these friendly young girls happily joined us at our table. They had a sweet and friendly attitude towards us, despite our lack of interest in going off with one of them.

Someone announced that the last bus was about to leave at some early morning hour. So, we piled in with the American sailors for the return journey to the docks. This ride was uncomfortable in a tropical downpour and the noisy, overcrowded bus. But I used my wet towel as a pillow on the seat in front of me and fell asleep.

When we finally arrived at the jetty, a loud party was in full swing under a thatched open-sided bar, filled with more American sailors. Since this was the last thing, I needed and as there were no available ferry boats in the continuing torrential rain, I headed back to the now-empty bus to lie across the rear bench seat and fell into a deep, exhausted sleep. When I eventually woke up, the sun was shining, and there was no one else around. Feeling confused, I wandered down to the jetty with no idea what I would do.

As I stood there, a lone US motorboat came chugging in to collect any stragglers. Being the only stranded person, they kindly took me back to HMS *Victorious,* where I climbed up one of the lowered boarding ladders and found my absence had gone unnoticed.

During my last days on HMS *Victorious,* we continued to operate from the ship as it ploughed its way through the Indian Ocean on its return run to the UK. I was fascinated when we cruised through the Suez Canal. It was unexpectedly beautiful, and what surprised me was just how close the banks of the canal were on each side of the ship's towering bulk.

As we continued west through the Mediterranean, my life on HMS *Victorious* ended. I did my last flight on the De Havilland Sea Vixen when the Squadron disembarked from abeam Malta. From there, we flew the planes back to RNS Yeovilton. There, our wives, and girlfriends, who were all prettily dressed in the warm August sunshine, welcomed us home. It was such a wonderful sight to see Tish, who had only recently flown back to the UK on an indulgence flight on an RAF Beverly transport aircraft. She now stood amongst the other wives and girlfriends, looking radiant and beautiful. We drove to Portsmouth's naval dockyard a few days later to collect my kit, where I took Tish on a tour around the vessel. It now looked incongruous and out of place in this very English setting.

Later, some of us met up, and I introduced Don McKenzie to my sister Antoinette, who was still in London and had recently qualified as a nursing sister. They immediately hit it off and got married very shortly after that. They had a son called Craig, and a few years later, they came to stay with us in Rhodesia after I had re-joined the Rhodesian Air Force. When I took Don to the pub at the Officers' Mess at Thornhill, he proved to be a real hit with his sparkling sense of humour. Sadly, soon after his return to the UK, he lost his life in a Sea Vixen that crashed short of the RAF Chivenor runway.

My mother later moved into a flat in Braunton, only 3 miles from Chivenor, which has now become a base for the Royal Marines, who operate helicopters from there. I recently learnt that the pilot of the Sea Vixen was Lt John Stutchbury, and they experienced some undercarriage problem. Then, while turning on short finals to land on an easterly runway, the Sea Vixen stalled and flicked straight into the ground. I was able to see the exact spot where the accident had occurred and had left a dent on the edge of a field in what they call 'The Downs'.

I know little of what became of other Fleet Air Arm contemporaries, except to mention one event. After a double engine failure, Allan Tarver received the George Cross for his bravery. The problem occurred when the observer's ejector seat failed to operate because something had jammed his solid overhead hatch. The only remaining option was to attempt a manual escape. Unfortunately, with both engines windmilling at such low RPM, there was insufficient hydraulic pressure to power the hydraulic flight controls. So, Allan triggered the emergency system, which popped a wind-driven hydraulic motor into the airstream and enabled him to continue flying the aircraft in a fast glide of over 220 knots, which he could only maintain at a high rate of descent. While still losing height, Allan repeatedly rolled the aircraft upside-down in several failed attempts to pop the observer out. But unfortunately, he had to eject seconds before the aircraft plunged into the sea, with the hapless observer still trapped inside.

Ian Kilgour, a Sea Vixen Mk 2 pilot, whom I met recently at a local church, told me that after Allan Tarver's incident, they designed a frangible plastic hatch, through which the observer could eject. Amazingly, we now both live near Shaftesbury, not that far from RNAS Yeovilton, which is now a helicopter base and has a well-established aircraft museum.

The Navy has a strange system with contracts, which recruits never sign. Queen's regulation stipulates that once a recruit accepts his first salary, he agrees to all the contract terms and conditions. Because I joined on a 16-year Medium Service Commission, the Navy refused to accept my resignation when I gave our first child's death as the motivating reason for wanting to leave. Instead, I was given a ground posting as the Hunter simulator instructor at RNS Brawdy in South Wales. This position was a more settled posting, and we leased a cottage on a small farm called Park Hall, which was close to one of the airfield's perimeter crash access gates. Here, I could cycle to work, lift my bike over the gate and then ride less than a hundred yards to the simulator and my office.

We spent a few happy months there before the Rhodesian Prime minister Ian Smith's Declaration of Independence, 11 November 1965. For months Tish and I had followed the lead-up to this declaration on the TV, and watched journalists interviewing Harold Wilson, the pipe-smoking British Prime Minister, and Ian Smith. Apart from his thick Rhodesian accent, I was immensely proud of Ian Smith. Following his declaration speech, four Navy captains interviewed me and asked what my reaction would be if tasked to drop a bomb on Rhodesia. I had to admit that I could never comply with this as I had friends and family there, and when asked what I would like to happen, I told them I would be happy to leave the Fleet Air Arm. Ten days later, I was returning to Rhodesia on the *Windsor Castle* passenger ship.

6 A SPECTACULAR TWINKLE ROLL

Hunter FGA9

I was flying for the Fleet Air Arm when Ian Smith, Rhodesia's Prime Minister, announced its Unilateral Declaration of Independence (UDI) from Britain. The next day, the station adjutant summoned me to meet with four Royal Navy captains to check where my loyalties lay. On stating that I would never enter warfare against Rhodesia, they asked me if I would prefer to leave the service to which I readily agreed, and we began to happily plan our return to family, friends, and the blue skies of Rhodesia.

I met Air Cdre Archie Wilson at Rhodesia House on the Strand in London. He was also leaving the UK and arranged for the consulate to book and pay for our return to Rhodesia onboard the *Windsor Castle* via Cape Town. I was informed that they would bear the cost if I worked for Air Traffic Control or the Air Force on my return. From Cape Town, we drove our

Austin Cambridge to Gatooma in Rhodesia, arriving without warning at Tish's parent's home, on Christmas Eve 1965.

In the new year, I started training to become an air traffic control officer at Salisbury Airport. However, my ambition was to qualify simultaneously for a commercial pilot's licence and return to flying as soon as possible. At that stage there was no possibility for me to rejoin the Rhodesian Air Force, as at the breakup of the Federation of Rhodesia and Nyasaland, the Air Force had declared they would not accept the return of anyone who had chosen to leave and join another service.

After six months with the ATC, I started to write the commercial licence exams. After the first day of writing, Sqn Ldr Shorty Dowden approached me and asked if I would like to rejoin the Air Force. After finishing the second day of the exams, I was invited to the Milton Buildings for a formal interview with the newly promoted Air Marshal Archie Wilson. I told him I would be delighted to rejoin the Air Force and asked if it would be possible to go back onto No 1 Squadron and fly the Hunter aircraft again. I also mentioned that my ambition was to become a PAI (Pilot Attack Instructor).

I had enjoyed flying the Hawker Hunter before I left the Royal Rhodesian Air Force. While in the Fleet Air Arm of the Royal Navy, I flew the down-rated Hunter FGA 11 version, which the Navy used as an advanced training aircraft. The Navy had also appointed me as the Hunter simulator instructor when I returned from an eight-month tour flying the Sea Vixens off HMS *Victorious*.

I had been one of the lucky few on No 1 Squadron, just over three years ago, when it had received twelve refurbished Hunter FGA9s. But most of the other pilots, apart from Rich Brand and Rob Gaunt, had since been transferred to other squadrons. It was still a pleasure to be reunited with the Hunter again. However, it was not that easy to slot back where

I had left off, especially as I had to forsake seniority and become junior to any officer who had elected to remain in the Air Force after I had left.

In the Fleet Air Arm, I had become used to a highly professional organisation, where the pressurised flying program ran like clockwork. They were also sticklers for the maintenance of accurate and disciplined battle formation positions. In contrast, the laid-back approach displayed by the younger pilots on the Squadron, was difficult for me to accept.

I had also gleaned some expertise during my flying in the Fleet Air Arm, especially when it came to accurate low-level navigation, for which I had learnt some valuable rule-of-thumb methods to make quick mental calculations while covering the ground at 420 knots (seven nautical miles per minute).

This knowledge helped me win a low-level navigation competition by a large margin, which No 4 Squadron had planned and set up. For this, they drew up twelve turning points that commenced from Thornhill Air Force base, where they positioned timekeepers. Halfway through the exercise, each pilot stopped for a break at New Sarum, before continuing with the second half which ended back at Thornhill.

All watches were synchronized, and every pilot was allocated take-off times, separated by ten minutes, and given the exact time to pass over each turning point. Just before startup, the pilot was handed a photograph of some feature to identify and pinpoint on one of the tracks.

Some of the turning points were on bush airstrips, and as each pilot turned onto the next heading, he had to decipher and record the message depicted by signal panels that had been laid out on the ground. By the application of my acquired navigational techniques, it was possible for me to fly over ten turning points within five seconds and two within seven seconds of the allocated arrival times. This accomplishment was in complete contrast with most of the other participants, some of whom even became totally lost.

Before my return to the Squadron, no one took any interest in aerial combat. But later, when the Air Force selected to put me on a (PAI) course, I took the opportunity to pass on some of the tips to other students. One of them was Steve Kesby, who was keen to learn the techniques I employed and became proficient himself.

However, I landed myself in hot water when I enticed other pilots to try a few new things. One of these times was when the Squadron deployed to Bulawayo Airport, where we lived in tented accommodation in a clearing to one side of the terminal building. On one formation sortie, I suggested to Dag Jones that we practice a run over the field in low-level 'battle formation' with a 200-metre separation between us. Once overhead, we would carry out a synchronized 'twinkle roll' before breaking into the circuit. A twinkle roll is a complete and rapid 360° roll around the longitudinal axis. Everything went well during our practice at altitude, where I called "Twinkle, twinkle, go!" We would then do a simultaneous 'twinkle' before entering a climbing turn to simulate the formation break onto the downwind leg, the standard practice used when a formation joined the circuit. However, when we put it into practice over the airfield, I was unaware that Dag had allowed his aircraft to drop dangerously low. The first time I knew that things had not gone off as planned, was when I taxied onto my parking spot, and Rob Gaunt, the A Flight Commander, came storming across the tarmac to give me an absolute rocket.

Back on the Squadron at Thornhill, we often took the opportunity to practice cloud formation in 'Guti' conditions, phenomenon that occurs in the Midlands when low clouds and drizzle is pushed in by high pressure rounding the Cape Peninsular. On one such day, Rob Gaunt suggested that Don Northcroft should take the opportunity to do some cloud formation flying, in these conditions. However, Don, who was busy on his OCU training and had recently received his wings and the *Sword of Honour* on his course, declined the offer, as he felt he needed a bit more practice in clear weather first.

The next day, a reported tropical cyclone was sitting off Beira in the Mozambique Channel. The civilian Met Officer, Harvey Quail, asked if two Hunters could be sent up for a closer inspection and report its position. I was assigned to the task, with Don Northcroft as my wingman.

We took off in clear blue skies, and after half an hour, approached this cyclone weather condition, which looked like a colossal atomic bomb, rising like a gigantic mushroom to an unbelievable height above us, and a flat skirt of lower white cloud below us. I was impressed by how miniature Don's aircraft looked against the cyclone's central towering column.

After a short time in the area, we headed back home. But to my surprise, although on the way out the countryside had been visible, there was now one solid thick cloud up to 25,000 feet, and ATC at Thornhill reported a 200-foot cloud base in heavy rain there. Having declined the offer to practice cloud formation the previous day, Don was now committed to formatting on me, all the way from the cloud tops, and then finishing off with a formation landing in a heavy downpour.

ATC instructed me to join overhead the field using the non-directional beacon (NDB) and carry out the standard timed teardrop NDB letdown. From there, the radar operator would guide us onto a GCA (a ground-controlled approach). The GCA controller would then call out corrections to our heading and rate of descent to keep us on the centerline and glide path.

Everything went well until I established us on the inbound turn. As I took a glance to my right to check how Don was doing, my heart skipped a beat. All I could see was thick dark cloud. My mind went into a spin, wondering how we could join up again, or if Don could complete the letdown on his own. Then, as this was all going through my mind, Don's aircraft reappeared. I breathed a sigh of relief as I realised that he had only been momentarily hidden from my view by slipping into the 'line astern' position. After that, Don managed to stay tucked into the 'echelon starboard' for the remainder of the approach. But although he had never practiced a formation landing in bad weather conditions before, it was now the real thing. As we crossed the threshold, I called, "Cut, cut!" and delayed throttling back for a couple of seconds before touching down on the water-logged runway.

Tol Janeke was the Acting Flight Commander while Rob Gaunt was on leave. He was obviously in complete panic when he noticed the rapidly deteriorating weather conditions and was concerned about Don's limited experience. When we taxied into the dispersal area, Tol came at me like a raging bull as I was climbing out of the cockpit. Despite being unaware of the rapidly deteriorating weather conditions, and that ATC had downplayed the heavy rain, he was sure it was negligence on my part that had put us in this predicament.

I had flown with Rob Gaunt when he was one of the more senior pilots on the Squadron before I joined the Fleet Air Arm. He was a likeable character who knew how to milk the system and arrange *'jollies'* away from the base. These gave pilots some extra perks, and I went on three interesting jollies before joining him on a different one to South Africa.

The first trip was a day stop to Beira, where during Easter or `Rhodes and Founders` in July, I had often hitchhiked to as a teenager. On those occasions, it could often take up to 12 hours to get there. But on this Hunter flight, my brain could hardly absorb that it could be the same place. From takeoff to landing, the trip took a mere 38 minutes.

Another trip I did was as Rob's number two when we spent a day at Chiredzi in the Lowveld, a low-lying area of

Rhodesia, full of citrus and sugar cane farms. By the time we were ready to return, Rob had organised 55 pockets of oranges and grapefruit for us to take back. We stacked these pockets in every conceivable space on the aircraft, including the 'Sabrinas', the bulging fairings that caught the spent 30mm cartridges and links. When we ran out of all available nooks and crannies, we just emptied the remaining pockets onto the cockpit floor around and behind the ejector seats. It made me shudder to consider what a board of enquiry would have made of their investigation if one of us had needed to eject.

My third trip was with Roy Morris to Lourenco Marques (now Maputo), Mozambique's capital city. At the plush Polana Hotel, I enjoyed the continental atmosphere and a prawn cocktail with lunch, while Roy went off to play golf with the Rhodesian ambassador. Shortly after lunch, I went sightseeing, but I felt sick with nausea and stomach cramps when I reached the fish market and only made it back to the hotel in time. I suffered from seafood poisoning, which was so violent that I thought I was dying. I could not imagine how I could fly back the next day, but fortunately, the ambassador recognised the symptoms and gave me some tablets, which resulted in a quick recovery. But ever since then, I have been allergic to shellfish.

I discovered through life that it is impossible to recapture water that had flowed under the bridge, which was noticeable when I bumped into previous girlfriends after rejoining the Air Force. Apart from having different pilots on the Squadron, I had become a different person myself. I was now married with a young son, and apart from gaining experience, I had seen another side of life and taken a few emotional knocks. In retrospect, with the hindsight of age, I realise that I had returned to the Air Force with an 'attitude', and what a dreadful pain I must have become. So, it did not help to win friends when Rob Gaunt selected me to join him on a trip to the Western Cape province of South

Africa. Ostensibly, the trip was to test a rain-repellent system, recently installed on two of the aircraft. The timing was not suitable for tests at home, as we were in the dry winter month of July.

Consequently, we needed to go south to search for rain, but I suspect the whole exercise was just an excuse to go on another 'jolly'. Some pilots, who felt more deserving because of their seniority and time on the Squadron, felt resentful when Rob selected me to accompany him on this trip.

We duly set off, after the thrill of drawing up the appropriate maps and working out the flight plans. The trip itself lasted two days, with a stopover at Pietersburg. With Rob as the leader, we took off in formation and were soon at altitude, doing nine nautical miles per minute in the clear crisp winter sky. After takeoff, I moved into the high-level battle formation, 1,000 meters off his beam. We crossed the Limpopo River's almost dry bed, and in no time, contacted Pietersburg Air Force base to join their circuit.

That night we were entertained by our South African hosts, and despite the South African Air Force mess regulation to close the bar by 1900 hours, they kept it open well into the early morning. Rob was friendly by nature and was more accustomed to handling this kind of entertainment. I am also not sure how they managed to get around the strict SAAF pub time. I just remember staggering back to our billet for a few hours' sleep, waking with a debilitating hangover and excruciating headache the next day.

After breakfast, an arrangement was made for the pair of Hunters to join and formate with a South African Air Force Sabre. We then carried out a fly-past over the airfield for some photographs. When the United Nations pilots had flown it, the Sabre had previously proved itself against the Russian Mig 15s, during the Korean War. But next to Rob's Hunter, which remained rock steady throughout the flight, the Sabre's instability looked awful. Once we had joined up on either side of it, I was able to see how unstable and chunky it looked, with its notorious snaking movement known as Dutch rolling, against the aesthetically clean lines of the Hunter.

We then set a course for the Langebaanweg Air Force base to the north of Cape Town. We flew at Flight Level 390 (39,000 feet) during the flight, which was way above all other scheduled commercial flights. As we raced over the unfamiliar terrain almost entirely devoid of indigenous trees, I noticed how different the Orange Free State's rolling grassland was from the more familiar Rhodesian woodlands. While following the Vaal River, which meandered below us, we caught sight of Kimberly's big holes, now full of dark blue water. They had been the diamond workings of the last century, where Cecil John Rhodes had made a fortune at the tender age of 24. Shortly after that, we crossed the Orange River and overflew the scrublands of the Karoo. After our first glimpse of the Cedarburg Mountains and Picketburg, we joined the Langebaanweg Air Force base's circuit.

The next day, when I met Rob before breakfast, I learnt that he was in extreme pain from haemorrhoids, making it impossible for him to sit down or walk at a reasonable pace. Rob's slow speed took nearly forty minutes to get from the Officers' Mess to the flight line, a distance that should have been a brisk ten-minute walk. So, it became my responsibility to carry out the trials for the experimental rain-repellent system. After a short briefing, I took off in search of clouds and rain.

After a quick visit to the met office, I headed south towards Cape Point, where I buzzed in and out of clouds without much success, before carrying out a letdown approach on DF Malan Airport. As I broke off my final landing approach on finals and was overshooting the runway to return to Langebaanweg, the tower controller called me on the radio saying, "Ysterplaat Air Force base has asked you to show them your feathers".

This base was not far from Table Bay Harbour, so I asked for their ATC frequency to carry out a flypast along their northerly runway. Once I picked up the runway, I dived from 10,000 feet with the throttle fully open. As I approached, I knew how the distinctive humming note emanating from the jet engine would thrill the spectators. An earsplitting roar would then follow this sound as the aircraft passed overhead, and an incredible crackling sound, like loud rustling leaves, would be left behind.

However, I failed to consider the increase in engine power produced at sea level, compared to over 4000 feet in Rhodesia. This increase in thrust caused the aircraft to accelerate to just below the speed of sound. The ATC had given me carte blanche license, and being a born show off, I flew far too low and sped like a bullet fifty feet above the runway to perform a twinkle roll abeam the tower, which must have looked spectacular from the ground.

The Hunter's hydraulic-operated controls are incredibly responsive to handle, allowing for an impressively fast roll rate when the pilot pushes hard to one side. The control input would spin the aircraft wings around its length while the aircraft continued to fly straight. It is such a rapid roll rate that one could almost say that it happened in the twinkling of an eye. Hence the name 'twinkle roll'.

As I had performed this manoeuvre many times without a problem, I never gave it a second thought. Fortunately, I raised the aircraft's nose above the horizon before its execution. I expected a quick snappy roll to occur but was surprised by the unexpected lack of response. Instead of the usual breathtaking rate-of-roll, which should have taken less than a second, the result was a terrifyingly drawn-out slow roll, while the nose dropped back to the level flight position.

Those on the ground saw a slow roll performed by an aircraft flying at a breakneck speed, while screaming low down the runway. As soon as I completed the roll, I pulled the aircraft's nose up as high as possible to get away from the ground. I was horrified by the unexpected lack of response, which took place in slow motion, while I held my breath. The experience shook me so much that I headed straight back to Langebaanweg, with my legs starting to jump uncontrollably in reaction to the surge of adrenalin.

The controller was so excited by the most foolhardy and spectacular bit of flying he had ever witnessed. He called back over the radio, "Everyone asks if you could come back and do that again!" But having only narrowly escaped the jaws of death, there was no way I was going to repeat that!

I never mentioned anything about my death-defying manoeuvre to Rob, and after much thought, I realised that this was due to the aircraft approaching the speed of sound. This faster-than-expected airspeed must have caused a shockwave to develop over the wings, which would have broken up the smooth airflow over the ailerons and resulted in their becoming less responsive.

At that stage, Ysterplaat was home to the SAAF navigators training school. Years later, I returned there to undergo a quick conversion onto the Alouette II, which would become used in our Force for initial helicopter training and make it possible to free up the Alouette IIIs for operational requirements. Now, as a helicopter instructor in the Rhodesian Air Force, I met one of the trainee navigator instructors, who had seen this manoeuvre.

Unaware that I had been the pilot involved, this navigation officer described how he had witnessed a Mirage III aircraft perform the most spectacular slow roll just above the runway. He remained unaware that it had been a Hawker Hunter, flown by an overconfident and impulsive young

GEORGE WRIGLEY

Rhodesian Air Force pilot.

7 LEOPARD AIR IN MALAWI

Cessna 337

After returning to Rhodesia from the Fleet Air Arm, I worked at Salisbury Airport as a trainee air traffic controller. But my heart was not placed in the job, as I pined to get back to flying, and at that time, there was no prospect of my being able to get back into the Air Force.

Operating out of Salisbury airport was a small charter company called RUAC, which flew Piper Apaches, and other light aircraft, used as charter aircraft for passengers to fly into small bush strips and neighbouring countries. My ambition was to get my commercial pilot's licence to apply to join them. With that in mind, I commenced studying for the forthcoming examinations of eight gruelling subjects set by the Department of Aviation. Meanwhile, Ting Orbell, the chief training captain with CAA (Central African Airways), invited me to come for a test flight on one of their DC-3 Dakotas. It must have gone quite well, apart from the fact that he was a bit horrified when I tried to haul the lumbering bus around in a 45-degree bank turn. I was unfamiliar with transport aircraft, as the last planes I had flown were Hawker Hunters and the Sea Vixens, both relatively high-performance jet fighters. After the flight, he informed me that as soon as I passed my written examinations, they would employ me as a trainee first officer and give me a conversion onto the DC-3.

In retrospect, it might have been better to have followed through with this offer. But, after completing the first day of examinations, S/Ldr Shorty Dowden approached me and invited me to rejoin the Rhodesian Air Force. I was so excited by this offer and the prospect of going back onto jet fighters that I could not concentrate on the examinations the next day. Because of that, I failed to pass one of the exams on this first attempt. Unfortunately, at that stage, I still craved to do aerobatics, air-to-air combat, and fire rockets, and cannon at the bombing range. So, flying a DC-3 seemed like driving a bus when all I wanted to do was drive a sports car. Later in life, when I did become an airline pilot, I found that it had its own challenges and demands, and the contact with the crew and passengers more than compensated for the apparent lack of excitement. But by then, I was a lot older and had become more mellow.

Once I became re-ensconced to the Air Force way of life, I managed to rewrite and pass the exam paper I had failed. Then with the generous help of F/Lt Pete McClurg, who kindly lent me £100 to take a test flight on a Piper Cherokee 140, I became one of the few Air Force pilots to have a commercial pilot's licence. Tish then kindly did some temporary work in Gwelo for one month to help us repay this debt.

In 1969, after three and a half years back in the Air Force, I was approached by S/Ldr Bob Woodward, who had somehow learnt that I possessed a commercial pilot's licence. He spoke to me about a friend in Malawi, who owned Leopard Air's aviation firm and desperately needed a pilot to fly the

Cessna 337 Skymaster. This plane is a twin-engine push-pull aircraft, which he leased to the Malawi Railways, and he needed a pilot for two weeks over Christmas and New Year, while his regular pilot took his annual leave. This opportunity was a windfall, as not only was it an opportunity to make a bit of extra money, but I could also take Tish to Malawi for a holiday. Fortunately, we could leave our two children, Shane, three years old, and Kim, nine months old, with Tish's parents in Gatooma before heading off.

Once we arrived in Malawi, we were settled into the Shire Highlands Hotel in Blantyre for the duration. The hotel was a wonderful experience, with a charming African flavour and drums to announce mealtimes. Here we feasted on the delicious bream fish (Cabo) from Lake Malawi, and while we were there, two mixed-race couples were the only other guests. It was an unusual sight for us at that time in our lives, as mixed marriages were still illegal in Rhodesia. In our allocated hotel room, we could hear African music pulsating from a bar beneath our first-floor balcony and watched South African registered cars pull up at all hours of the night. From there, they would pick up ladies of the night, with their tight miniskirts straining over voluptuously rounded behinds.

One of the mixed couples was older and more dignified, and the wife was quite exotic, with light skin and aquiline features. I surmised that she might have come from Ethiopia. The first time I noticed the young couple was while I was looking at some postcards in the foyer. I saw the tall blond husband as he arrived at the reception desk. He had a posh English accent, which I would have called a Pommie accent in those days, and amused me when he asked the receptionist, "What are you having for Christmas nosh?" His beautiful young African wife followed him into the foyer. Her dress was a full-length flowing flamboyant caftan, which was fashionable at that time. There was still a hint of a neat trim figure beneath the shroud of soft material and the loose fit.

We never conversed with these other two couples, apart from courteous nods and sat at tables spread apart in the dining room, where we spoke in muted tones in the subdued dark atmosphere. We were served by friendly Malawian waiters with their broad smiles and white teeth.

One afternoon we were treated to a remarkable display of an approaching 'Chiperone', a local weather phenomenon caused by high pressure driving warm moist air from the Indian Ocean and up the mountainous area in that part of Malawi. Churchill Road climbs a hill to the left of the hotel, and one warm, sunny afternoon, as we sat on our balcony, we felt the wind picking up. This freshening wind was followed shortly by the appearance of a thick black fog, that swirled down the road towards us. Within the space of ten minutes, a low dark cloud engulfed us. It became so gloomy that cars needed to use their headlights with visibility below 50 metres. Shortly after that, we were driven indoors by fat heavy raindrops pelting down on us from the blanket of fog. I had never witnessed anything like that before and did not realise that this same weather would catch me out on a return trip to Chileka Airport, within a few days.

Despite doing a flight test, writing the Malawi Air Law paper, and doing a technical written exam on the aircraft, it only permitted me to fly VMC (in visual meteorological conditions). Although I was a qualified instrument examining officer in the Air Force, I did not have an instrument rating on my commercial licence. To get one would have necessitated another written examination, which I had not felt worth the extra trouble.

Three days later, the lousy weather forecast did not deter me from boldly flying my first flight out of Chileka Airport to take the Railway salaries to Monkey Bay and Salima. These airfields are on Lake Malawi's shore, and I had to fly through thick clouds to get there. However, with the arrogance of youth and the confidence gained after years of instrument flying, I did not let a small thing like not having a rating on my commercial licence bother me.

However, the only thing that precluded my landing at either place was when the air traffic control recalled me, as the Railway office had phoned them to say that rain had flooded both grass runways.

The Company then tasked me to collect two passengers from Lilongwe, Malawi's new capital, which had only been built recently, in the middle of the African bush. The Railway department requested that these passengers be transported to Blantyre and brought back later.

This made it possible for me to take Tish with me to spend the day with her cousin Joan who lived in Lilongwe. I would collect her when I brought the passengers back.

After dropping Tish off, I duly returned with two passengers to Chileka Airport. On the way back, the air traffic control advised me that the airfield was clamped in by a Chiperoni, with the cloud base down to 100 feet and the visibility less than 200 metres. Despite having witnessed this weather phenomenon from the safety of our hotel balcony only a few days before, being a 'steely-eyed' jet fighter pilot, I was not about to be put off by a bit of low cloud. So, I pulled out the bulky Malawi Air Navigation Regulations manual and opened it at the let-down plate for Chileka Airport. At that time, the only let-down facility available was an NDB (non-directional beacon). On studying the let-down plate, I noticed a prominent mountain feature off to one side of the runway. As we flew, I examined the let-down procedure for a while and then asked the passenger next to me to keep the manual open on his lap, so that I could glance across at it.

This awkward letdown procedure did not guarantee an accurate lineup or distance to the runway. I missed the break-off altitude and continued the dangerous descent towards the ground. When I looked up at about 50 feet above the surface,

I saw what looked like a road, appearing out of the thick low cloud and drizzle. This sight startled me until I realised it was the runway threshold. I somehow managed to pull the aircraft around in a tight left-hand circuit while keeping visual contact with the ground. With the aircraft wing barely 50 feet clear of the ground, I managed to land safely on the runway. When I taxied onto the dispersal area, a young man came running to talk to me. He was another pilot, and the terrible weather had delayed his departure, and he asked me how I had managed to land in those conditions. I later learnt that the weather had precluded a BOAC aircraft and another commercial airliner from approaching to land. They had to hold off for another hour before making a safe approach and landing.

Later that day, I returned with one passenger to Lilongwe, thinking nothing of the fact that there was now only one person when I had expected two. I undertook the return flight that afternoon in comparatively good weather conditions. I was able to retrieve Tish, who was entirely oblivious to my previous irresponsible behaviour and had spent an enjoyable day with her cousin.

Later that evening, back at the hotel, I was approached by a gentleman who I did not recognise as the person who had been the other passenger in the back seat that morning. It was apparent from his slurred speech that he had been busy imbibing copious amounts of alcohol at the pub. He then apologised, "Sorry, I didn't come back with you, but I didn't enjoy flying in all that mist!" I later learnt that the Railway department had brought him to Blantyre for a disciplinary hearing.

The rest of our stay in Malawi was enjoyable. Chip Kay, the owner of Leopard Air, invited us to join him, his girlfriend, and two young daughters for a Christmas lunch at his lovely home near Zomba and looked after us well. One day, he took us to Luchenza flying club, where I met Ralph Casey, our course commander, when I had been an officer cadet seven years before. He had also left the RRAF at the Federation's breakup, and I always felt he had been extremely unfair and dished out unnecessarily harsh punishments. After receiving my wings, I held a deep-seated resentment towards him. So, when I saw him, I gleefully thought that this was my opportunity to let him know just how I felt. With clenched teeth, as I approached him, I said menacingly, "Hello Ralph!" but the wind was whipped out of my sails when he responded with a vague, "Don't I know you from somewhere?" which was a good lesson to me in the futility of nurturing a grudge.

That same day, a dashing and irresponsible private pilot took Tish up for a ride in a Piper Cub. However, his flashy and reckless display of bravado angered me. Although it seemed to thrill the other spectators as he buzzed around the field, all I could focus on was Tish's long blond hair whipping out of the open-air cockpit.

While staying with Chip Kay that Christmas, I came down with laryngitis and could hardly talk. The next day he asked if I could join him to fly a single-engine Piper Comanche from his grass strip on a plateau, overlooking the Shire Valley. The grass had not been cut for some time and was so long that it was higher than the wings' leading edges. This grass created an extra drag force on the plane, and the wings collected many grass seeds. It was fortunate that they had perched this airfield on top of a plateau because the aircraft managed to stagger into the air just before it reached the end of the runway. The flat-topped escarpment allowed us to descend enough to build up speed before climbing away.

But just as we approached the edge of the plateau, I coughed loudly. It sounded like a loud metallic click, which gave Chip a real fright as he thought that the engine was about to give up the ghost. Until that moment, he had been the epitome of a colonial gentleman, with precise, clipped speech, a neat moustache, and a permanent bow tie. I could not help having a perverse chuckle to see him momentarily losing it.

On our last day in Malawi, once again the torrential rain was bucketing down in Blantyre, and before we left the hotel, I was in the foyer when our posh young English gentleman was checking out. He was busy carrying his suitcases to the car, which he had parked at the back of the hotel.

As he stepped into the soaking rain, his pretty young black wife appeared, shouting out to him, "*Duhling havie yoo gotie my sundoos?*" With his beautiful Etonian accent, the posh young voice replied, "Whaat?" To which she responded in some frustration, "*Havie yoo gotie my Sundoos, my Choos?*" Her accent so tickled me that ever since then, whenever Tish has looked for her sandals or shoes, I have responded, "*Duhling,* are you looking for your *choos*?"

8 JOYS AND TRIALS OF A FLYING INSTRUCTOR

As a jet fighter pilot during my early years, my overriding ambition was to become a PAI (pilot attack instructor). The Fleet Air Arm calls this qualification as being an AWI (air weapons instructor). These titles held a quixotic ring of flamboyance and flair about them, and this sat well with my self-image as a steely-eyed jet fighter pilot.

After two years in the Fleet Air Arm, I re-joined the RRAF in July 1966, and then enjoyed a stint on Hunters, before being sent on a PAI course at Thornhill Air Force Base. F/Lt Pete McClurg conducted this course, which was carried out in the side-by-side seated Vampire T11 and the single-seat version, the FB9.

Of course, only the British would build two versions of the same aircraft, where one has the radio transmit button on the throttle, and the firing button on the control column, while the other has them the opposite way round. As to be expected, this mix-up often led to a few incidents on the inbound turn at the firing range. When pilots tried to transmit "Turning in live," they inadvertently fired the weapon into the bush instead.

I enjoyed the sporty look and feel of the FB9. Its small round nose, which the pilot could not see from the cockpit, made the compact windscreen the only thing in front of him. I also enjoyed the 'spade grip' control column, which added to it feeling like a little sports car. The controls were lighter, making the FB9 nippier than the lumbering T11 with its protruding bulbous nose. The only drawback was that unlike the T11, the FB9 did not have an ejector seat. So, pilots had to strap to and then sit on the packed parachute as a cushion, like we did on the piston Provost.

In an emergency, the lack of an ejector seat made it difficult to escape the FB9. The recommended method was to trim the nose fully down, wind back the canopy, roll inverted, twist, and press the 'quick release box' to free the seat harness, and then let go of the control column. In theory, the pilot would pop out of the cockpit like a champagne cork to avoid striking the aircraft's tail. However, Bruce McKerron, a well-liked young pilot, was tragically killed in the general flying area near Selukwe, after an inadvertent engine flame-out, which could have happened in a stall-turn manoeuvre. With no suitable terrain within gliding distance for a forced landing, he had no option but to bail out. However, the board of enquiry discovered that during Bruce's attempt to escape, he must have struck something that broke his arm. Sadly, this would have made it impossible for him to pull the parachute's ripcord. This discovery led them to speculate that his arm had probably been struck by the tailplane at the end of the twin booms.

Although never discussed, FB9 pilots generally accepted that the chance of surviving a bailout from an FB9 flying low level was very slim. I remember my heart missing a beat on one sortie, when I had rolled inverted for a climb away from a stint of low flying. I remained upside down far too long for the small submersible tank. This tank allowed for about 20 seconds in inverted flight, which I must have exceeded, as the engine suffered a partial flame-out. I forgot all about pressing the relight button as I hurriedly rolled the aircraft the right way up and pulled the throttle back. To my relief, the engine managed to spool up again.

As far as I was concerned, it made no difference whether I flew in the Vampire T11 with its ejector seat, or the FB9, as there was little chance that I could have made a safe escape from either. When I was a cadet, because of my height, there was concern that in the T11, my knees could hit the top frame of the windshield if I had to eject. So, a team carried out some tests with me sitting in the aircraft. After pulling the ejector seat handle, they winched it up with a block and tackle. This handle pulled a cable and its attached blind over the top of my bone dome. But thankfully, with all the attention focused on my knees, no one noticed that because my neck was too long, the cable exited the seat from below the top of my bone dome. So, instead of the blind holding my head safely back against the headrest, I had to nod it forward to operate the cable. Because my head was now bent forward, my neck would have broken in an ejection. Later, when I flew the Hunters and Sea Vixens, this problem fell away, as both aircraft had alternative handles situated in front of the seat pans between the pilots' thighs. This ejector handle, which was pulled upward, gave me a safer option if I needed to eject.

The PAI course covered carrying out low and high dive-bombing, firing 3-inch rockets, 20 mm cannons and doing air-to-air interceptions. A camera mounted above a first-generation computerised gyroscopic gunsight filmed these interceptions. Filming would start as soon as the pilot pressed the firing mechanism, and in theory, rounds fired by the attacking aircraft would hit the enemy aircraft when the central bead was on the target.

This gunsight was developed in the later stages of World War II. But we found them difficult to operate while hauling your aircraft in a curved attack. To get good air-to-air results, the pilot had to keep the *'pipper'* (the aiming dot of the gyro-sight) on the enemy aircraft, and simultaneously use the throttle grip to adjust the distances between the inside tips of the six surrounding diamonds to make them coincide with the target's rapidly expanding size between the wingtips as we closed in. Operating the gunsight in this way would compute the distance between the two aircraft and calculate how far the guns would have to lead ahead. But follow-up assessments of the films showed disappointing results.

However, this exercise was very academic, as the target

had to fly in a straight line at constant airspeed, which would be highly unlikely in actual combat situations. Even then, with conditions set up so much in favour of the attacking pilot, a successful attack was easier said than done. It made us appreciate the difficulties World War II pilots experienced, and further heightened our respect for them when we realise how grateful they would have been to receive these improved gyroscopic sights on their squadrons.

The whole course stimulated me as it was right up my street. To check my proficiency, Wg Cdr Mick McLaren, then OC Flying at Thornhill and later became the Air Force Commander, took me for my final check ride. The trip was a simulated introduction to using the 20 mm cannon against ground targets at the Kutanga bombing range near Que Que. As we got airborne and the undercarriage started to retract, without any forethought, I casually asked him what he would do if the guns began to fire at that moment. He replied, "I don't know. What would you do?" In a matter-of-fact way, and without trying to be clever or cute, I continued, "Just put the wheels down again, as there is a micro-switch in the landing gear, which cuts off the circuit to the firing mechanism." Somehow, this seemed to impress him. I then demonstrated the attack pattern at the range and let him fire at one of the allocated hessian targets, and the whole flight went off without a hitch.

During the debrief afterwards, he casually informed me that he would be putting my name forward for the next FIS (Flying Instructor School). His suggestion floored me, as it was not my ambition or plan to become a general flying instructor. I blurted out, "But I'm far too young!" Regardless, undeterred, and true to his word, the Air Force transferred me to No 6 Squadron, the flight training squadron, within a few months.

Dag Jones joined me on the same FIS to become AFS (Advanced Flying School) instructor on the Vampire T11. We completed the course just in time to instruct the students of Number 21 PTC, who had just finished BFS (Basic Flying School) on the Piston Provost. Bernie Graaf and Ed Potterton were my first two students and Ed ended up winning the Sword of Honour as the best all-around student in academics and flying.

When I started to instruct these two, I was a harsh taskmaster, but they responded well to my demands. But by the end of the course, the flight commander Gordon Wright, felt I had become too arrogant. So, from the next group, to bring me down a peg or two, he decided to allocate me three students who were not over-endowed with much natural ability. Meanwhile, Dag and I had undergone further training at the FIS, to qualify as instructors for the Basic Flying School to train raw students on the piston-engine Provost. After completing the FIS, Dag and I were ready to instruct students from Number 22 PTC. Gordon allocated three students for me to teach. They were John Blythe-Wood, Dick Paxton, and Roger Watt. Each of these students, for various reasons, was not deemed to be a promising prospect but went on to prove themselves brilliant operational pilots. John had his own game-capturing helicopter business, and Roger earned himself the Silver Cross, the second-highest award for valour in the Rhodesian bush war.

I kicked off in the same uncompromising manner I had previously adopted. But unexpectedly, after introducing circuit flying, I started to tirade a sensitive student during the debrief session and was shocked when he burst into tears. I had not previously experienced this in my military flying career and was so embarrassed, I gruffly muttered, "You had better go outside and come back when you have stopped blubbing!" As he left the office, I realised that I would need to adopt an entirely different approach to my style of instructing.

Making the appropriate change was not easy for me, as I was not acutely tolerant towards any form of slapdash flying, or sympathetic towards students who were slow to learn. My most significant difficulty was when students struggled with instrument flying and took unbearably long to notice that a wing had dropped and the aircraft was slowly drifting off the required heading, or if the nose dropped or climbed above the horizon and resulted in a departure from their height. I always wanted to scream the moment a change went unnoticed because of inattention to the AH (artificial horizon).

If I did the instrument scan for him, the student could never progress. So, I had to accept that the student's 'scan' would not improve if I continued to draw his attention the moment anything changed. So, in the end, the only way I could stop myself from drawing their notice as soon as things started to go wrong was to bite my lip and stare out of the window. Nevertheless, this did not help me relax. After a while, I developed what I now know as an irritable bowel syndrome. Not realising what caused this, I went to our Swedish camp doctor, who had some weird ideas on natural remedies and recommended eating yoghurt. Yuk. The yoghurt we ate at that time was long before Greek or fruit-flavoured yoghurt had come onto the market. But nothing I tried seemed to help. So, in the end he sent me to the Bulawayo General Hospital for a 'barium meal.' But all this proved was that there was nothing physically wrong with me. In retrospect, I understand that my symptom was due to the stress caused by my intolerant attitude when giving instruction.

One day on the squadron as a Provost instructor, the flight commander asked me to take an aircraft up for a test flight. With the Navy brass' tolerant attitude still fresh in my mind, I never gave it a second thought when I proceeded to do an aerobatic display for Tish and our two-year-old son over the farmhouse we were renting. The property was situated 500 feet below the Ridge, just outside the downwind leg of the main tar runway's circuit pattern. Unfortunately, W/Cdr Chris Dams, the new OC Flying, happened to be looking out of his window and he caught sight of my display. I can understand how surprised he must have been, when he saw an aircraft pop into view at the top of each loop, and then vanish below the Ridge again, before momentarily reappearing.

As I climbed out of the cockpit in the dispersal area, the flight commander told me that the OC Flying wanted me to report directly to his office. Unfortunately, he had already informed the newly appointed Station Commander, G/Cpt Frank Mussell, who in turn had already phoned Air Force HQ, in Salisbury, to ask what they should do about this unauthorised breach of discipline.

I think HQ would have been happier if they had dealt with it at the station level. But once HQ had been put in the picture, there was no choice but to deal with the matter themselves. A few days later, I was flown in a Vampire T11 to New Sarum Air Force base, driven to HQ on Jameson Avenue, and marched without cap and belt to appear before the Deputy Air Force Commander, AVM Archie Wilson. He calmly told me that he was awarding me a reprimand and a six-month seniority loss. I then returned to Thornhill to continue with my usual flight duties.

After instructing on one more course, HQ posted me to the helicopter squadron. The irritable bowel syndrome became an immediate thing of the past, even after I became an instructor on the Alouette for nearly six years. It was somehow different. Especially as I enjoyed teaching qualified pilots on such an exciting and versatile aircraft. It had so many different roles, that I found helicopter flying genuinely fascinating.

I now feel sorry for the students who had to endure my initial instructional technique. But in the end, I became one of the few pilots to qualify as instructors, and Instrument Rating Examiners, on both fixed-wing and helicopters. My ambition had never stretched beyond becoming a Pilot Attack

Instructor, but the Air Force never used me in this capacity despite being trained for it.

Once I left the Air Force and went into commercial flying, I did a bit of instruction on helicopters but never pursued being an instructor again.

9 NIGHT SORTIE AT WANKIE

While I was on the No. 6 Training Squadron at Thornhill, the flight commander asked me to re-familiarise F/O Dave Becks on the Provost fixed-wing trainer. He had been one of the first pilots posted to the newly established No 7 Squadron with the Alouette III helicopters' procurement. One night, he had a mishap while carrying out a hovering exercise just above the water at Lake McIlwaine. The water was so calm that the pilot-controlled landing light showed up at the shallow bottom of the lake. I am unsure of the circumstances, but judging his height above the surface wasn't easy.

Before he realised it, the helicopter had dipped into the water, which immediately smashed up the tail rotor blades, and the Alouette went out of control. Dave had no choice but to cut the engine and abandon the helicopter as it began to sink, and I recall having seen some photographs of the tricky salvage operation.

Dave was a gentle person who lost his nerve after this incident. Because of this, he was taken off all flying duties and trained as an ATC officer in the control tower, where he excelled.

He later decided to put this traumatic experience behind him and requested a return to flying duties. The Air Force agreed to give him every opportunity to accomplish this. They suggested some training should take place away from Thornhill's busy and pressurised environment. They chose Wankie Airport, as it would be a more peaceful and undisturbed location. The site was considered especially suitable for night flying, which had been the central cause of concern.

They tasked me to reintroduce Dave to night-flying,

and we duly set off for Wankie the following day in a Provost. I experienced no problem with Dave's flying ability during the positioning flight. After a few circuits at Wankie, we had lunch at the forward airfield (FAF 1). After this, we had a short break and then briefed for the night flying training, which we planned to start just before sunset and continue into full dark.

Everything went well until we received a radio call from the FAF Commander, who informed us that a No 5 Squadron Canberra had failed to return from a night flying navigation exercise. One of the turning points was over Victoria Falls, and a game warden at Robin's Camp in the Wankie Game Reserve, reported that he had heard a loud explosion not far from their base. Dave and I were asked to discontinue the night flying training and go and check out that area.

After a short while we saw a veld fire in the area, which led us straight to the crash site. We then circled overhead for a few hours, hoping that the crew had managed to eject safely and would give some indication to that effect. In the bright moonlight, I made out a distinctive long *vlei*, running in a southerly direction, which I was sure I would be able to identify the next day. After a while, as our fuel was running low, we called off the search, and returned to Wankie.

Early the following day, a Dakota transport aircraft arrived, and the crew asked me to take them back to the spot where I had found the aircraft wreck. Fortunately, I managed to relocate the long *vlei*, and within a short time, both crew members, Jim Stagman and Dave Postance, were seen alive and well and HQ dispatched a helicopter to uplift them.

The crew had both ejected from the Canberra, but Dave had parachuted down some distance from his pilot,

Jim. Instead of Dave showing any gratitude that we had found the crash site the night before, he merely complained that our flying overhead was more of a nuisance than any help and said that we had disturbed a pride of lions, which were then being driven towards his position, and made it difficult for him to get a decent night's sleep! I am not sure why neither of them tried to signal the overflying aircraft with their night-flying torches.

Although trained as an air force navigator, Dave Postance, later went on to study for his commercial pilot's licence and joined Air Rhodesia as a pilot. I believe he later went on to fly For Trans Australian Airlines, which later became QANTAS.

It was expedient that we had been in the area and assisted in the search and rescue operation. But sadly, Dave Becks was not comfortable with the extended time we had spent flying that night. The flight proved to him, once and for all, that he had not overcome his apprehension towards night flying, and as far as I know, he never returned to flying duties again.

10 NAVIGATION FLYING – TRIBULATION TO TRIUMPH

It was many years after I began my flying career, that I realised that I was not blessed with a good sense of direction and became quickly disorientated when walking in the bush with no visible reference points. Though scary at times, it seemed understandable and quite normal. To this day, I still become disorientated in big shopping malls, mainly when the architect had not designed them to be built at right angles but go off in all different directions.

I often turn the wrong way when I leave a building or car park, and after years of exasperation, I have realised that more in line with the opposite sex, the way I find my way around is by going from one recognised reference point to the next. I have also realised that, when travelling by road, I do not keep a mental picture of every twist and turn along the way, and most of the time, I have no idea where north or south is. A little while ago, I lived in a mountainous area in the Eastern Districts of Zimbabwe. To get to where we were living, we often drove on the main road approaching Juliasdale from the southwest. It follows several hairpin bends and even doubles right back on itself a few times as it descends to the bottom and then climbs out from the sides of the canyon. We could see Manyoli Mountain, situated behind Pine Tree Inn, first appearing straight ahead when approaching this canyon. It then moved abeam, and then later behind us, and as we continued, I could then see it in front again. This cycle repeated itself several times, and it took a few months before I realised that it was the same mountain that I was seeing each time.

When I first drove around the UK, I would become regularly lost. In Salisbury (Harare) and Gwelo, where I spent my early years, the developers had laid out both cities in neat square blocks, with most roads travelling at right angles to each other. These designs are in total contrast to the towns and cities in Europe, which have been developed over thousands of years and radiate from several central points, and seldom have any right-angled intersections. With this confusing network of roads in built-up areas, and before the advent of GPS and Satnav systems, I used to navigate by reference to various pub signs and other noticeable shops and buildings.

This system worked well enough for me. But more recently, after an absence of twenty-three years, Tish and I made a return visit to Yeovil, near to where we had lived when I was training on Sea Vixens in the Fleet Air Arm. We had spent the first four months of our married life in a nearby quaint Somerset village. Every Saturday, we would go to a Chinese restaurant and then to a local cinema. When we returned on holiday, we tried to find this restaurant to reminisce on our early married life. However, what we saw was considerable development that had taken place and there was now a mind-boggling network of flyovers carrying busy lanes of traffic around and overhead the city. It was so confusing that I knew how Rip Van Winkle must have felt when he came back from sleep after twenty years! It was so disorientating that we could not even start to imagine where our little Chinese restaurant might have been.

One can imagine that this disability can be a real disadvantage in flying, especially in the early days of aviation in aircraft with limited navigational aids.

During basic flying training, on my first solo night cross-country on the piston-engine Provost, my instructor sent me to fly a simple route with only two turning points, from Thornhill to Mashaba, then to Filabusi, and return.

When it came to the second turning point at Filabusi,

the town was so small and insignificant that it did not even register for me that so few lights could be worth considering. Instead, the bright lights of Bulawayo, which I saw straight ahead on the horizon, attracted my attention. Somehow, I forgot everything my instructor had taught me about maintaining an accurate heading and time. By the time I realised my mistake and set course for base, I had to call the ATC controller for a 'QDM', the code for the correct heading to steer. In my innocence, I was oblivious to the panic and concern I had caused my instructor and ATC by being so long overdue. Fortunately, I made a safe landing on the grass runway, lit up by gooseneck flares, with some relief. When the Squadron technicians had drained the aircraft fuel tanks, they told me that there was four gallons of fuel left, which would have kept me airborne for only a few more minutes!

As my flying career progressed, it became necessary to overcompensate and discipline myself to maintain steady headings, speeds, and times. I was also careful to pick out distinguishable features to confirm that I was setting off in the right direction. This practice was easy enough when flying Vampires and Hunters in Rhodesia, which had a limited number of human-made features. But when I started flying in the UK on the Hunter conversion at RAF Chivenor, and later flying Hunters from RNS Brawdy in South Wales, I was quickly confused by the cluttered spider webs of rivers, roads, and railways.

To overcome this difficulty, I concentrated on accurate flying and adopted the quick rule of thumb calculations, that was taught in the Fleet Air Arm. The system enabled me to check on my progress by selecting only very distinctive features three minutes apart. If it was an exercise demanding a specific arrival time, I applied a formula based on the number of nautical miles the aircraft flew in one minute. At 420 knots, the plane covered seven miles per minute, and I would adjust the speed by twice the number of seconds out at the checkpoint. I then held this difference in speed for three and a half minutes, half the number of miles per minute, before making a slight correction to account for the head or tailwind component. This is only possible in aircraft where the pilot could make significant speed changes.

If I was flying in an aircraft at 360 knots, I would hold the change of indicated airspeed for three minutes.

However, during low-level navigational exercises, the effect of the surface wind on fighter jets flying at high speed over the ground, is usually negligible. Also, any crosswind component that could cause the aircraft to drift only required minor adjustments to the aircraft's heading for it to remain on track.

When I returned to the Rhodesian Air Force after two years in the Fleet Air Arm, I had become proficient at this type of navigation, and managed to win a low-level competition by a considerable margin. Before the exercise began, we first synchronised all our watches. We then flew on a pre-programmed route, which included twelve turning points. Our take-off times were staggered, and pre-positioned observers were waiting on the ground at each turning point. They knew which 'call sign' should pass overhead at any given time and would jot any discrepancies down to the nearest second. Before start-up, some team members would hand us photographs of features we had to look for and pinpoint on some part of the route. There were also message panels laid out on the ground at some turning points, which we had to jot down and decipher as we passed overhead at 420 knots.

As it turned out, a few of the pilots failed to see some of the turning points, while others got utterly lost. However, by adhering to the previously described technique, I managed to arrive overhead ten out of the twelve turning points within five seconds of the allocated time. I was only seven seconds off the required time at the other two points. It proved that I could reach a high standard in navigation, despite having a lousy sense of direction.

Unfortunately, these tricks of the trade proved to be inadequate when HQ transferred me onto helicopters. These Alouettes were devoid of any navigational aids with simple and marginally useful compasses. But due to their slow speed across the ground, they were also severely affected by the wind. One advantage, which many other pilots and I had to make use of at one time or another, was the option and ability to land near a farmer's house and ask for the name of the farm.

I spent many hours dividing my one-in-a-million map that displayed Rhodesia, to show a grid depicting all the one-in-two-fifty-thousand series. I further divided these maps into grids, to display the larger one-in-fifty thousand series. With this arrangement, one could carry out a long-distance flight at a comfortable height, using the one-in-a-million map that spanned the whole country. Then, as we got closer to the destination, we could transfer to the more detailed one-in-two-hundred-and-fifty-thousand map at some easily recognisable feature. For a less distinguishable position or precise detail on the ground, we would need to move onto the smaller-scaled one-in-fifty-thousand map, which we used when in support of search and rescue or military operations,

For coverage of the whole country, we often carried no less than 96 of these smaller-scale maps, which had to be organised in such a way as to make it possible to find the correct one quickly and efficiently. It helped us avoid lengthy searches for the correct map and lose track of the activity on the ground.

The most challenging situation was when the destination was in the adjoining corners of several maps. Anyone who has ever had to use map reading to navigate in the air, or while driving on unknown roads, will sympathise with this dilemma. These days many people have never had to bother studying or even carry a map as they have become wholly reliant on their Satnav systems. It has made for more relaxed trips and possibly even decreased the divorce rate.

By the time counter-insurgency operations in Rhodesia had moved into full gear, I had organised the different scale maps into sets. These I kept in specially designed canvas bags with Velcro-secured flaps that the Parachute Section had made up for me and were admired by members of other military units that operated with us. In due course, the other helicopter pilots also adopted this system.

Permission to fly the Alouette was only allowed when there was a visible horizon. The only flying instrument that showed the helicopter's attitude to the natural horizon was the artificial horizon (AH). Although it was not a fundamentally flawed device, its pneumatic source that provided suction to spin the instrument's gyro was prone to failure, because the stainless-steel venturi was attached to the sidewall of the jet pipe. This weakness made it vulnerable to a fragment of ejected carbon, which could clog the venturi's constricted waist that provided the required suction.

The other instrument operated by this suction was the directional indicator (DI), which is also prone to drift off heading on each flight. Pilots needed to continually realign it with a small magnetic compass hanging from the Perspex windshield's centre frame. This compass was in an inverted cup-shaped dome, suspended in alcohol to dampen oscillations. It displayed the aircraft heading in increments of 10 degrees, with the actual numbers at 30-degree intervals. It was not easy to get an accurate reading from this instrument at night. Especially as this small compass was awkward to read and not well-lit. All-in-all, it made the Alouette a real navigational challenge at night!

Under conditions of low visibility and no visible horizon, in their zeal to provide backup and support to the troops on the ground, many pilots felt obliged to continue flying. Unfortunately, the ground forces took for granted that pilots could fly under any weather condition, and those who stuck to the regulations would unfairly earn the reputation of being "Chicken".

One night, while I was on deployment at Mount Darwin in the northeast of the country, a truck full of troops came under attack in an ambush near Marymount Mission. I led three Alouette helicopters delegated to ferry doctors in, and take casualties out, of the ambush position. There, I went into a clearing lit up by a small fire and a few torches. On the ground, I witnessed a scene of carnage while the other two helicopters followed behind me, with Mike Litson landing next. The third pilot to land was Neil Liddel, a South African pilot on detachment to Rhodesia. I became concerned when I noticed his difficulty judging his approach under the pitch-black and hazy conditions.

Once on the ground, the accompanying two South African doctors examined the wounded individuals lying on the side of the dirt road. The first casualty had suffered a severe head wound and made awful animal sounds. After making a quick assessment, the doctors moved on to the next victim, ignoring Mike Litson's protestations. When the doctors completed what they could do, they suggested flying the most critically wounded soldiers straight back to the Salisbury General Hospital. Realising that the conditions were way below the minimum specified in our operations manual and concerned by Neil's struggle on his landing approach, I suggested that he should return to Mount Darwin. Then Mike and I flew the two critical casualties to Salisbury General Hospital.

This task proved to be one of the worst flights I had ever done in a helicopter. The haze was so bad that I could not see any visible lights or horizon for the first fifty minutes. After that protracted time, the lights of the small rural town of

Shamva, appeared from behind some scattered hills.

The whole flight took over one-and-a-half hours, and the casualty with a terrible head wound travelled on a stretcher behind me. He continued to emit eerily inhuman moaning sounds the entire way. His condition was considered hopeless, as his brain damage was so bad that the doctors felt he would have to spend the rest of his life in a vegetative state if he survived. Surprisingly, although they had given him minimal attention, he continued to live a further twenty-four hours.

In retrospect, I struggled to overcome my inherent terrible sense of direction. But since then, I have passionately believed that people can excel in their areas of greatest weakness. Achieving this is most likely due to the extra effort required to compensate. I certainly gave a lot of attention to my map reading and took much pride when I eventually achieved a high standard. I also took delight in pinpointing little details shown on the map, which most of the other pilots ignored. Especially after visiting the map department, where I saw what meticulous and loving care the topographers used deciphering the aerial photographs that were used to draw up the maps. After hours of monotonous filming runs by Canberra crews, the photographic department sent the rolls of film to them. I think it would have pleased them to know how much someone enjoyed locating the most obscure features, such as the depicted cattle dip tanks.

There were certain areas in Rhodesia, like parts of the Zambezi Valley, Wankie, and the Lowveld, where the terrain is so flat and devoid of habitation that there is little to go on. However, after flying in an area for a while, I learnt to distinguish the slightest rise of ground or recognise individual trees with distinctive features, like a dead tree or one of slightly different colour or height. I would often use these features to pinpoint me in a vast, featureless landscape covered by Mopane trees.

Despite the significant improvement in my navigational skills, I would still have days when I lost concentration and could yet become disorientated.

However, once I started flying as an airline captain, the flights were tightly controlled by ATC and followed designated departure routes and airways. In later years, I progressed onto more sophisticated aircraft with continually advanced navigational aids. But at times, even these aids were prone to failure, with some disastrous consequences that I describe in other stories. Before I retired, GPS navigational systems made the need for map reading a thing of the past.

11 START OF HELICOPTER FLYING IN 'THE STICKS'

Alouette III

Before serious terrorist incursions into Rhodesia began in earnest, defence force units occupied a few base camps near the North and Northeast borders. One of these base camps was called Rushinga, near the Ruya River, a tributary to the Zambezi River. The Army located the base camp on top of a small oval-shaped hill, which had a good vantage of the surrounding countryside.

After completing my Alouette III conversion, my first tour of duty was a week at Rushinga base, which had a helicopter pad tucked in between corrugated iron buildings.

Each day, the camp commander utilised the helicopter to drop a 'stick' of troops at pre-planned positions. After being put down, they would patrol a given route, looking for tell-tale signs which might indicate a group of terrorists entering or leaving the area. Each week, they also used the helicopter to change the relay teams operating from a high point on the Mavhuradonha mountain range, situated to the northwest of Rushinga, not far south of the small border Mukumbura settlement. These teams' task was to relay radio messages between various patrolling units and the base camp ops room.

During my first few tours of duty at Rushinga, the routine became boring and predictable after dropping off the stick, until some flight engineers persuaded me to carry out minor diversions on our way back. Some suggested we fly along the Ruya River to look for wild game to supplement our supply of rations with some venison.

One day on our return, we saw a troop of baboons darting across a dry riverbed. The poor terrified animals scrambled away in sheer panic, while looking back over their shoulders. A few of them clutched their young ones to their chests while they scampered over the granite rocks. The flight engineer, who had managed to scrounge a few extra rounds of ammunition for his FN Rifle, begged me to let him have a potshot at them.

I orbited at about 300 feet above the ground while Billy went into a killing frenzy on rapid fire. In no time at all, I was sickened to the stomach when I noticed the red bottom of one animal, as it scampered between some rocks. Not realising that it was the natural colour of their buttocks, I thought it had been hit by one of the shots. This sight horrified me as he continued to fire wildly in every direction, while I felt he should rather finish off, what I thought was a wounded animal. Not wishing to continue as an accessory to this carnage, I pulled away, saying we had better get back to base. I automatically assumed that Billy would feel as sickened as I was. But I was stunned by his obvious thrill and enthusiastic expressions of gratitude. Until then, I had never witnessed such a killer instinct, and felt ashamed to contribute to this wanton slaughter. Apart from specifically hunting for the pot, I refused to indulge in any such senseless activity again. But later, when using a mounted MAG in running battles against terrorist groups, I learnt how difficult it was to hit a moving target from a helicopter. So, I imagine the chance of hitting one of those fleeing baboons with a handheld rifle would have been negligible.

On a later trip, we were asked by the camp commander if he could come on one of these sorties to shoot something for the pot. After searching for some time, we came across a magnificent kudu bull. As this beautiful creature loped gracefully through the bush, I moved alongside it in a fast 'air taxi', just skimming the Mopani treetops. Fortunately, there was an extra passenger sitting right behind me, who watched straight ahead while I concentrated on keeping alongside this majestic animal. The passenger's wild tapping on my shoulder drew my attention to the branches of a baobab tree, that stretched up like a dry hand above the general tree line, waiting to pluck us out of the air. In horror, I yanked up on the collective, and adrenalin coursed through my veins, as providence spared the life of another beautiful creature. Once more, I had a lucky escape, and learnt an invaluable lesson.

A few years later, a helicopter operating out of Rushinga went missing. The pilot was John Smart and the flight engineer, was Tinker Smithdorff. After an extensive search, a fixed-wing pilot eventually found the helicopter wreck, far from their authorised flying area, and sadly the accident had killed both crew members. Although I never said anything at the time, I sometimes wondered if an unauthorised activity, like my near miss, was the reason for this tragic loss.

When anti-terrorist operations began in earnest, we had less opportunities to participate in such irresponsible behaviour. Unfortunately, helicopters are so versatile, that pilots soon start to enjoy their newfound freedom, which offers them more options to lead them astray.

When one is young, it is easy to fall into the trap of turning into a 'bush pilot'. This adopted frame of mind, makes it difficult to re-adapt to the disciplines required when transferred back to flying fixed-wing aircraft. Especially the strict parameters required when operating a fast-flying jet aircraft.

On one of my last trips to Rushinga, a territorial company occupied the camp. As I climbed out of the helicopter on my arrival, I immediately recognised Graham Vaughn, who was the officer in charge of the unit. I remembered him well from my time at Prince Edward School, where he had been a few years ahead of me, and I had admired for his athletic ability.

Later, a request came through to exchange two separate teams, operating individual relay stations, from the top of the Mavhuradhona Mountain. They were from different RLI Commandos, based on the mountain's northern side and just off the road leading to Mukumbura. As there were no passengers between Rushinga and the relay station, I invited the 'cooks and bottle washers' along for the ride. Some of these men had never been in a helicopter before, as they spent all their time confined to the base camp, where their duties restricted them to catering and vehicle maintenance. Graham also volunteered to join the others in place of my flight engineer. I accepted his offer, as I knew I could manage to get in and out of the spacious landing areas unassisted.

It was a real treat for these passengers, who could hardly contain their excitement as we flew directly to the relay station on the first leg of the sortie. When I landed there, they disembarked, and enjoyed the chance to look around while I continued with the two trips to change over each team.

When I had fulfilled the task, I re-collected our passengers. Once airborne, we scudded across massive mountain-acacia treetops that surrounded the relay station. As we left the mountain plateau, the ground fell sharply down a cliff face, where I held level flight for a moment. Then, to add to their thrill, I surreptitiously 'dumped' the collective into autorotation, a 'minimum pitch' position, which pilots continually practiced in preparation for an engine failure, and made the helicopter drop like a brick.

After clearing the top of Mavhuradonha Mountain, I allowed the helicopter to plummet towards the ground passing through 3,000 feet in just over a minute. I then raised the nose in a flare to level off just above tree height. During the level off, I was aware that one of the passengers directly behind me, had said something to Graham. So, once I had re-set the collective to hold 20 feet above the treetops in forward flight, I asked Graham over the intercom what they had been discussing. He informed me that the man sitting directly behind me at the open door had been terrified. The whole way down, he had gripped the leather handle above his head so tightly that his knuckles had turned white. Then, as we levelled off just above the trees, he asked, "Was that an air pocket Sir?"

When Graham told me what this passenger had said, I had to climb the helicopter a few hundred feet, as I could hardly fly the aircraft as I began to cry with laughter.

Although I only wanted to fly jet fighter aircraft, helicopters' versatility and personal contact with the ground forces compensated for this change in my flying career.

12 TWO WEEKS IN CHICOA-MOZAMBIQUE

Flying Alouette from Chicoa

As 1972 was coming to an end, I spent Christmas and New Year in Chicoa, a small Mozambique hamlet marked on maps dating back as far as 1640. It was a settlement on the Zambezi River's southern bank and upstream of Tete, which the Cabora Bassa dam's rising water has submerged since then.

I was part of a detachment of four Alouette III helicopters of No 7 Squadron, which was deployed to assist the Portuguese in their struggle against the Frelimo insurgents. Intelligence indicated that Frelimo was assisting ZANLA in their operations northeast of Rhodesia. The Portuguese were also being harassed by increasing land mines and ambush incidents. These were predominantly along the road between Chicoa and the small Rhodesian border town of Mukumbura. This increased activity prompted the Rhodesian authorities to volunteer helicopters, tracking units, and backup troops from the RLI, to assist the Portuguese armed forces in conducting more effective follow-up operations.

Our contingent slept in tents alongside a house allocated as our ops room, next to the village football pitch from which the helicopters operated. Unfortunately, our accommodation was downwind of the *aldeamento* (protected village), into which the Portuguese had placed a burgeoning population removed from their rural villages to stop them from assisting the Frelimo terrorists. They now suffered appalling squalor conditions and lacked proper hygiene or adequate toilet facilities. As the prevailing wind was from the east, escaping the stench wafting towards our camp was impossible. This breeze resulted in a constant horde of flies that buzzed annoyingly around us, and where they settled on everything. Each person had to carry a homemade fly swatter around to combat them, and at mealtimes we had to continually wave our knives and forks over our plates to stop the flies from settling on the food.

Chicoa had a raised water tank to provide for the town's needs. But apparently, this was never brought into service, and the only time the tank held water, was after the Cabora Bassa dam submerged the whole village! The only available water to the inhabitants was muddy-brown, having been collected in old fuel drums from the Zambezi River banks when we were there. We used this water to wash our clothes and to shower under a small drum hanging from a nearby tree. Although this countered the heat and humidity, I continually felt grubby, with grains of sand remaining in my hair. One day, during a heavy downpour, many of us took the opportunity to shower naked in the rain. The soap and shampoo left us feeling squeaky clean for a change.

The detachment had been in place for two weeks before I arrived with a replacement crew. Sadly, we had to spend both Christmas and New Year away from home.

W/Cdr Ossie Penton had been sent out with the first crew to establish an ops room and liaise with the local Portuguese commander. Both Ossie and the colonel were of small stature and formed an immediate *entente cordiale*, which resulted in some amusing anecdotes that blossomed out of their relationship.

Without a doubt, the Frelimo intelligence became aware that the Rhodesian forces had come to help the Portuguese counterinsurgency operations, and during the first two weeks, occurrences of land mine detonations and ambushes dried up completely. The absence of any activity left the tracking teams and helicopter crews hanging around in frustration. Then one day, as we were relaxing on the ops room veranda, entertaining ourselves in idle banter with Wng Cmdr Ossie, in typical crew room fashion, we noticed the Portuguese colonel come bounding across the football pitch. On reaching us in a triumphant mood, he exclaimed, "Good news, we've been ambushed!"

The colonel's next visit followed an accidental discharge from a Portuguese helicopter gunship 20 mm cannon. The gunner had detached it from its mounting to clean it on a trestle table next to the aircraft, and without first making it safe, he fired off a remaining live round in the barrel. This projectile struck the centre pole of the Portuguese cook's tent, which sent a piece of shrapnel into the unsuspecting chef's buttocks. A flurry of excitement around the camp culminated in the chef's dramatic evacuation to the Tete Hospital.

We were eager to hear the Colonels's reaction to this breach in safety procedures and what disciplinary action he would take against the gunner for such negligent conduct. There was no doubt in our minds that one of us would have faced a court-martial for a similar offence. So, when someone asked what would happen to the gunner, we were astounded by the colonel's casual attitude. He just shrugged his shoulders in a typical continental manner and, with a smile, declared, "You know, boys will be boys? *Sheet Cook* anyway!"

On the first night in Chicoa, the helicopter crews shared a communal tent with a local Portuguese contingent. Sometime after midnight, everyone was jolted awake by a blood-curdling scream. Lying there, frozen on camp stretchers with thoughts of an impending massacre and hearts pounding in their ears, they strained for other unusual sounds in the surrounding darkness. Out of the dark, a Portuguese voice suggested, "I think he was having a *night horse*". The tension evaporated and put everyone into fits of relieved laughter.

During much of our time in Chicoa, we battled with the flies and the oppressive heat. Stripped to the waist while sitting in the shade of the ops room veranda, we played endless rounds of Bridge. In the meantime, our two excellent caterers, both as odd as 'two-bob notes', kept us amused with their camp behaviour and unexpected quips. While the outside temperature soared to 45 degrees Celsius, one of them practised his yoga exercises in the blazing sun, dressed only in a loincloth.

Our Portuguese hosts kept our gas-operated deep freezers stocked with Pepsi Cola and continental-tasting orange drinks. However, I found that neither of these drinks could slake my constant thirst. Then in desperation, I tried an ordinary cup of tea and some heat fatigue tablets, which immediately worked wonders. Ever since then, I have sworn by the recuperative powers of this simple beverage.

With their flamboyant style, the Portuguese were an endless source of amusement to us. Each day, a few soldiers of the '*Páraquedista*' (paratrooper) Regiment swaggered across the football field *en route* to the squalid *aldeamento*. They were always immaculately attired in full camouflage regalia, including polished boots and grey berets, which they pulled down in dashing style above one eye. They would re-emerge a short while later, proudly flaunting giggling village girls, hanging onto their arms as their trophies. Their chests would puff out even further as they invoked a cacophony of wolf whistles and cheers from our camp.

Our pilots and engineers had good morale as we all worked together to keep the helicopters serviced and clean. Bob Mackie brought along his portable radio, which he played all day. We soon learnt the words of the current hit tunes and joined in loud accompaniment to Neil Diamond's 'Cracklin' Rosie' and 'I can see clearly now the rain has gone', by Creedence Clearwater Revival.

One day, a Portuguese soldier visited our camp with a baby vervet monkey. Our visitor hardly had a chance to settle down before his pet jumped onto one of our pilots, Barry Roberts, who was sitting on the veranda without a shirt. A thick mat of black hair covering the back of his chest and shoulders, made up for the sparse covering on his head. With a barrelled-shaped torso which earned him the nickname "Barrel," he continually played the clown, and kept us amused for hours. Whenever anyone commented on the lack of hair on his head, he would quip "I am trying to lose weight. Who wants fat hair anyway!".

To everyone's amusement, he now sat unperturbed as we ragged him while the baby monkey, that felt united to his surrogate mother, nuzzled, and blissfully suckled into Barry's chest hairs. We were then sent into more hysterics when the vervet went berserk, bared its sharp little teeth, and defiantly clung onto Barry, when its owner got up to leave and tried to take it back.

We were also entertained and often appalled by the Portuguese helicopter pilots' lack of airmanship. Their flashy style indicated a total disregard for the aircraft limitations in the Pilot's Operating Handbook. In contrast, we religiously adhered to these limitations drawn up by the Aerospatiale test pilots. One Portuguese pilot always arrived with a flourish, when despite the Alouette being restricted to 18 knots in a sideways air-taxi manoeuvre, would always fly sideways along the entire runway length above 40 knots. This excessive side-on force would have subjected the tail boom to tremendous stress. Then to crown it all, after landing he would always climb out of his helicopter, with the rotor blades rotating at full RPM and no one at the flight controls. This would always cause our pilots to watch in horror, as we realised that it put the helicopter in danger of being flipped over by a gust of wind.

Occasionally, we combined with Portuguese helicopters in follow-up operations, which quickly deteriorated into a complete nightmare. We would join over a preselected spot but never had a chance to meet up with them for a debrief, as they always returned to their base in Tete. Apart from the language problem, the Portuguese pilots never stopped shouting over the operational frequency. Their constant chatter made it impossible for our leader to give directions to the Rhodesian formation. As a standard safety procedure over the operation area, we always flew our helicopters in left-hand orbits. This ensured that we circled in the same direction and aimed the mounted MAGs towards the ground so they could then be brought into immediate action if required. But apparently, this was of no consideration to the Portuguese formation, which flew in all directions over the target area and resulted in total chaos.

Apart from this, when the Rhodesian pilots had to land in the same clearing, we would always follow one behind the other. The lead helicopter would land as far forward as possible, and then to minimise any chance of collision, we would take off in the same strict order,

However, on one joint operation, we joined a Portuguese formation flying Aerospatiale Puma helicopters, which were larger than our Alouettes. I led four helicopters to drop off our troops before the Portuguese Pumas brought in their soldiers. Then, just as I was about to lift off, a Puma flew directly over me and landed straight in front of my take-off path. Not only could this have resulted in a mid-air collision, but a blinding cloud of dust also smothered us, and his downwash buffeted my smaller craft. Once again, there was no opportunity to discuss this breach of airmanship with the pilot involved.

Shortly after that, we flew to the Portuguese side of the border town of Mukumbura. Our task was to arrive at first light to uplift a crack Portuguese Commando unit section. However, when we arrived, we saw no Portuguese troops ready for uplift, and shortly after landing, a bleary-eyed commanding officer came out to meet us.

With the rank of captain, he was a rough character with rotten teeth, and as an air force flight lieutenant, I was his equivalent rank, so he referred to me as "Captain". As a delaying tactic and to appease us, he invited us in for cups of coffee, which gave his motley troops time to ready themselves, as our idea of first light was a foreign concept to them.

As we sipped strong black coffee, I learnt that he was from the island of Madeira, and he enthusiastically vindicated their delay as the result of a fantastic party they had enjoyed the previous night. He unabashedly continued to expand on the night's frivolities and invited us to spend a night with them stating, "You can meet all the beautiful Rhodesian girls who came to the party last night. They speak good English too!"

We often spotted warthog and antelope during our missions. So, when we finished all our fresh meat, our new camp commander, Wng Cmdr John Mussell, authorised me to take a few passengers to go in search of something for the pot. Joining us on board was an army colonel, plus George, the Portuguese translator born in Lourenco Marques (now Maputo), and our ops officer Bruce Smith. Bruce was a qualified Air Force pilot who had stopped flying and became attached to Thornhill's Pilots' Ground Training School.

We flew off to areas where I had previously spotted some wild game, but as luck would have it, there was nothing to be seen. After an hour of fruitless search, we caught sight of a flock of Egyptian geese and decided one of these would have to do. As we drew near to them, they jinxed and swerved from side to side until one finally broke rank and plunged like an arrow, with folded flat wings, and dived straight into the swiftly flowing Zambezi River.

Hovering the helicopter alongside the goose, which the swiftly flowing Zambezi River swept along, allowed the Colonel to shoot it with his FN rifle, at point-blank range. The problem then became how to retrieve it. The current carried the bird so fast that after a few failed attempts by my engineer, leaning out as far as possible from the helicopter step, we realised it would be an impossible task. George volunteered to dive in and pass it back up to the aircraft. He stripped down to his underpants and plunged into the swirling brown water, grabbed hold of the slain bird, and handed it to us. Then our problems began when we tried to get him back into the helicopter himself. Bruce offered an outstretched hand from the rear-facing front seat near the door, but as he stretched his leg backwards for balance, it pushed down on the collective lever. The further he leaned, the harder he pushed, and this pressure forced the helicopter to descend and dunk the left wheel into the water. As I fought to keep the aircraft up, I kept losing sight of George, who was whisked along by the fast-flowing current. In the end, completely exhausted, he managed to grab hold of the nose wheel.

With my heart in my mouth, I managed to drag him across the river, through thick reeds and deposit him on a small sandy island. I then gently lowered the helicopter and hovered beside him, with the left wheel just touching the sand. This enabled him to climb back on board. Worn out and disgruntled, he looked like a drowned rat as he slumped onto the rear bench seat. He had completely lost all sense of humour, whereas the rest of us, fully appreciating how narrowly we had averted a catastrophe, could not restrain ourselves from howling with laughter when we looked at him,

The Wing Commander was disappointed with our measly offering on our return to base. Still, we decided not to divulge the risks we had taken or how near we had come to losing both our Portuguese translator and the helicopter.

The next day our ops officer sent us to collect some of our troops who had been on patrol for four days. When we arrived overhead their position, thickly wooded Mopani trees rising 200 feet above the ground covered the whole area. For miles around us, we could not see anywhere to land, while in their frustration, the troops had expended 400 rounds of .762 mm ammunition, into one tree to no avail. Sadly, this effort made it painfully evident that it would be impossible to clear an LZ like that, and our only choice was to return to base for a two-man bandsaw, which we dropped through the thick canopy of trees, to the exasperated troops below.

We then returned to base and asked them to call us when they were ready. The troops on the ground called us back after struggling for a further two hours to cut a sufficiently large enough clearing for us to get down to them. Even then, we had to descend the 200 feet in a slow hover while moving from side to side, forward and back, to avoid clipping overhanging branches with our rotor blades. There were still too many stumps in the way, which made it impossible to land the helicopter safely. So, the troops had to climb on board from a low hover. The troop of men clambering on board in this fashion caused the helicopter to swing dangerously, from side to side. Once they were safely in my helicopter, we had to go through the same slow procedure to climb out of the LZ. The troop commander, 2nd Lt Johnny Dawson, was a handsome young man one seldom saw without his black hair slicked neatly down. But that day I was tickled by his dishevelled and pitiful appearance.

During the rest of our stay, follow-up operations proved to be a disaster. Whenever tracks led through a rural village, the Portuguese soldiers disappeared one after the other into various thatched huts, with sheepishly giggling village girls in tow. The Rhodesian trackers would then emerge on the other side of the village without any of the backup Portuguese. Furthermore, there seemed to be an understanding between the Portuguese and Frelimo to stop short of serious contact. As most Portuguese military personnel were on national service from Metropolitan Portugal, their overriding ambition was to survive the two-year call-up period, or some minor self-inflicted war wound would have them repatriated.

This standoff situation added to the frustration felt by our trackers, as whenever they got close enough to surprise the Frelimo insurgents, the Portuguese troops would loudly tap their weapon magazines or toss a hand grenade into a nearby bush. These warnings of imminent approach gave the enemy sufficient time to make a quick getaway, while leaving *sadza* pots on a roaring fire and their kit abandoned around their campsite.

Towards the end of our detachment, an unfortunate incident occurred as we prepared to take off from the football field. As the troops settled into each helicopter, there was the familiar whining sound of jet engines and rotor blades winding up, which was followed by a loud explosion. My immediate thought was that we had come under a mortar attack. I then thought there might have been an accidental discharge or that a hand grenade had exploded, and I dreaded the prospect of serious injuries. But as it transpired, a waterproof engine cover, which flight engineers fitted on each helicopter night, had somehow snagged in the rotor blade control mechanism on one of the aircraft. This occurrence resulted in extensive damage to one of the Alouettes and a subsequent court-martial found it was due to negligence by the pilot and flight engineer. The board of inquiry decided that they had most likely failed to remove the cover during their pre-flight checks. Fortunately, no one had suffered any injury.

Until this mishap, we maintained a high state of morale despite the heat, myriad of flies and other frustrations.

I was particularly impressed by the professionalism displayed by senior flight engineers, Bob Mackie and Simon Maitland, who always remained cheerful and continually set a standard of exceptionally high work ethics. I will also never forget this operation for the loud accompaniment to Bob's portable radio, the jokes, and continual laughter.

Over that Christmas period, while we chased the elusive Frelimo, the country had a rude awakening with the murder of the Kleynhaans family on Altena Farm near the Zambezi escarpment, in the rich Centenary Block.

Following that farm attack, things would never be quite the same again. This caused security forces to initiating Op Hurricane at the Centenary police station. However, the gradually progressive escalation of conflict, spread like a cancerous growth throughout the whole country over the next eight years. This eventually terminated life as we had known it in Rhodesia, and with birth of Zimbabwe, the subsequent decline of standards in every field.

13 LESLEY SULLIVAN FLIES TO MUKKERS

Security forces created Op Hurricane in response to the Kleynhans farm murder north of Salisbury, in the Centenary farming block. This was a prosperous tobacco-growing area, with the farm close to the Zambezi Escarpment.

The head of the security forces set up a Joint Operations Committee (JOC), at the BSAP camp near Centenary, with an RLI Commando, supported by Air Force Alouette helicopters, a Provost, a Trojan, and a Police Reserve Air Wing (PRAW) light aircraft.

The Kleynhans murder took the military intelligence entirely off guard, as the sparse daily incident reports misrepresented the extent to which steady incursions had gone ahead, and the local villagers had not reported the undetected build-up of activity in the surrounding Tribal Trust Lands.

In no time at all, it became necessary to set up a second JOC at Mount Darwin, the next town to the east of Centenary and the District Commissioner's office. Initially, this town had a gravel runway to the southeast. But after a few years, a contracting company constructed a tar runway to the west of the built-up area. This strong airfield made it possible for the DC3 and other aircraft to assist in the operation.

Another RLI Commando established its base alongside the town buildings, where the Commanding Officer, Lt-Col Dave Parker, put his headquarters and mess tents.

Initially, the Air Force allocated one Alouette III to operate from the football field near the town. Then each night, for added protection, the pilot would reposition his helicopter onto the lawn outside the front of the District Commissioner's Office. A few buildings surrounded it in this position and gave it some protection against a possible attack.

Early each morning, and well before the office staff came to work, the pilot would return the helicopter to the football field. This early repositioning of the helicopter avoided disrupting and disturbing the public by blasting them with dust each time the helicopter took off.

The army engineers erected several tents to the west of the town to accommodate the operations, including a radio room, officers' mess tent, and pub for the JOC and its staff. The ops tent had the usual display of maps, with pins depicting a history of reported incidences, and buzzed with the sound of operations staff and the signal team, talking on radios and field telephones. On entering this area, one could hear distant voices squawking and crackling over the HF radios, as daily 'situation reports' (sitreps) and orders were transmitted and received. Other radio calls were sent and received via the relay stations established north of Mt. Darwin, on the Mavuradhona range. These relay stations made it possible to keep in touch with troops on the ground and observation points (OPs) out in the field.

One of the signal staff officers was Lesley Sullivan, a well-known radio announcer, who had an early morning programme on the Rhodesian Broadcasting Commission. He was a cheerful character called up for duty in the signals tent. When his stint of duty was over, he would enjoy socialising in the pub until the early hours of the morning before he staggered off to his bed.

Each morning, when he shuffled his way to the signals tent, he was just in time to see the helicopter pilot start up and reposition his aircraft on the football field. In the beginning, it would remain there for most of the day in readiness for the day's requirements. He saw this procedure repeated like clockwork, day after day. So, one morning as the pilot Mark Mclean, was busy winding up the rotor blades for a lift-off, Lesley spontaneously decided to treat himself to a short flip across to the football field before starting his duty. So, unannounced, and uninvited, he ran towards the front of the helicopter, and imitating the 'stick leaders' he had previously watched, he gave a thumbs-up signal to the unsuspecting pilot. Receiving a returned thumbs-up from the surprised pilot, Lesley ran in like a veteran *'troopie'*, crouching forward to keep well clear of the rotor blades. He approached the helicopter straight towards the pilot and then jumped onto the rear bench seat.

When Mark had been briefed to position to the north of Mount Darwin, to the small border town of Mukumbura, where the army unit based there would use for the rest of the day, no one mentioned to Mark that another passenger would accompany him. However, this did not unduly surprise him. In the meantime, Lesley Sullivan sat back happily, still expecting the helicopter to circle the town and land back on the football field, from where he could disembark and continue to the Ops tent.

When the other ops room staff came on duty, no one knew Lesley's whereabouts or why he had failed to report for duty.

It was the first time Lesley Sullivan had been in a helicopter or ventured into the operational area. But he got a kick out of being in the middle of Africa and so far from civilization. In this dry and dusty border post town, he spent the rest of the day wandering around and befriending everyone. He certainly made the most of this break from his daily and mundane routine in the signals tent.

After his return at the end of the day, when he walked into the mess tent that evening, he was suitably repentant, understandably embarrassed, sheepish, and very apologetic. However, from that day on, his day at the small village of Mukumbura was immortalised in his early morning radio shows, with his broadcasts peppered continuously with references to *"Mukkers by the sea"*, the nickname, which the troops had cynically given the dry and dusty village, and could not have been further from the truth as it was hot, dry, and dusty, and more than 500 miles from the nearest coastal town in Mozambique, which would have taken more than 12 hours to get to by road.

14 NEAR CATASTROPHE AT MAZOE

During my six years as a helicopter instructor on No 7 Squadron, one task was familiarising students with the planning and procedures for going in and out of police landing zones (LZs) at night. This exercise was fraught with dangers, and it was my job to warn, and bring to their attention, the potential hazards to avoid. This training would be combined with day and night navigation trips around the country, where the instructors would fly with their students during daylight hours to show them where the police had positioned the landing pads in vicinity of their police stations. These landing pads were usually on a cleared piece of ground with a sizeable, white-washed circle, and an 'H' painted in the centre.

We would also show the students how to orientate the police station with other buildings in the vicinity, plus any power lines, telephone wires, aerials, trees, high ground, and other prominent features in the immediate area. The Squadron also kept a box of cards for many police stations around the country. Whenever pilots needed to fly to their LZs, they could prepare themselves by studying the card's details, which often included an aerial photograph highlighting surrounding hazards and other salient features.

Whenever we planned to go in and out of a police LZ at night, we would contact the station for permission to use their landing pad and request assistance. They were always happy to oblige as it gave them a chance to refresh themselves on how to lay out the torches for night landings. They would place two sets of six torches in the shape of a 'T'; laying one set on their back to shine straight up and then lean the second set in such a way that they shone towards the safest line of descent.

We would call the police station on their radio

frequency to advise them of our expected arrival time (ETA) when approaching the area. They would give us the local weather conditions, wind direction, and the approach heading during this call.

On our arrival overhead the police station at night, we would fly at a safe height to clear the vicinity's highest point. We would then look for the lights of the landing T, and once we spotted it, the helicopter would be flown directly over it to note the approach direction. Simultaneously, the pilot would orientate the landing zone to the other lights in the area, which he would use in the circuit to reference the landing T position, which could not be seen in the circuit and only appear quite late on the final approach.

We generally had two instructors on the Squadron. When one of the instructors felt his student had reached proficiency in a particular exercise, he would arrange a check ride with the other one.

At the time of the following incident, John Annan was the second instructor, and his student Bob Jones, who had only recently joined No 7 Squadron, was now ready for a check ride before being sent solo on a night navigation exercise. The solo check ride would always include landings at a few selected police LZs, which I did for Bob's flight test. But when the flight was over, I wondered whether he was competent enough to fly solo the next night. Although not able to put my finger on the exact problem, I could not shake off my nagging doubts. Most students posted to No 7 Squadron got a kick out of flying helicopters, which novelty, and versatility they enjoyed. However, I never felt Bob had his heart set on helicopter flying, and he lacked enthusiasm and confidence. Feeling as I did, I thought I should take one more ride with him to put my mind at rest. So, I decided to accompany him on the next sortie.

So, in the briefing before departure, I explained that I would not treat this as a further check ride but only come along as an observer. I explained this arrangement to both Finn Marcussen, the flight engineer, and Bob, that for all intense and purposes, they should ignore my presence, make all their own decisions, and relate to each other, as if they were on their own.

The first two arrivals and departures at Norton and Darwendale police stations posed no real problems. However, I noticed that Bob was hesitant and became disorientated very quickly, as he continued to Mazoe.

When we flew overhead the LZ, we managed to see the torches. But they immediately disappeared behind trees as we entered the circuit. Losing contact with these torches made it difficult for Bob to establish an accurate approach direction. He also had a problem orientating the police station with the other lights of Mazoe. Unsure of himself, Bob continued to fly around for some time to relocate the T. After a few circles overhead, Bob turned the aircraft onto an assumed downwind leg. But on turning back towards the police station, he was unable to relocate the torches, which were hidden behind buildings or surrounding trees, as he continued to approach from the wrong direction. He was also unsure where to find the police station among the other town lights, which exacerbated his confusion.

After several aborted attempts, I became so tense and irritated that I wanted to shout at him and tell him where he was going wrong. But at the same time, I resolved to stick to my decision to let him work it out for himself. In my struggle to keep out of it, I had to bite my lip and look away in complete despair, as it was becoming such a fiasco.

A fitted landing light on the Alouette could be swivelled up and down by the pilot. However, this was usually only switched on during the last few hundred metres of the approach, which the flying pilot would use to show up trees and other obstacles around the landing zone. In addition to this, the flight engineer used a handheld lamp with a cable plugged into a nearby socket, to allow him to watch the tail rotor and avoid other obstacles as the helicopter descended into the LZ.

When Bob was on his fourth approach, we came from yet another direction as he descended the helicopter, once again at 500 feet per minute and 65 knots. The only outside lights, shining from the helicopter at this stage, were the aircraft's navigation lights, which consisted of a green light on the starboard side, a red light on the port side, a white light at the tail, and two external red rotating beacon, one on top and the other on the bottom of the fuselage, which were there for safety reasons, to abide by the legal requirement for all aircraft flying at night.

After a while, I became so frustrated that I disengaged myself from this complete hash up, and just sat there fuming and stared blankly out of the front windscreen. It was a brilliantly clear night, and without really focusing, I became vaguely aware that the lights of the small town in front of us started to disappear from my view. The reason did not initially register, though I felt uneasy that something was not quite right. Then suddenly, the flight engineer shouted, "Gomo!!" A Shona word for a hill or mountain, often used by the security forces.

When I heard that word, I was immediately aware of our situation and realised that we were about to fly into the side of a hill. Motivated by instinctive self-preservation, a conditioned reflex action kicked in without conscious thought. I snatched the flight controls from Bob, and yanked the collective lever to the full-up position while pulling the cyclic back as hard as I could. This input pulled the helicopter into a high nose-up attitude, which converted the forward speed to zero to gain as much height as possible. At the same time, I was aware of a high-pitched squeal coming from overstressing the main rotor gearbox. While all this was happening in slow motion, I held my breath, as I anticipated an impending crash, and wondered what our chances of survival would be. Then, with great relief, I saw the lights of Mazoe start to reappear again as we staggered over the top of the hill, which we cleared by a cat's whisker.

Slowly exhaling, I continued to climb to a safe height. Then the after-effect surge of adrenalin kicked in and turned my knees to jelly and uncontrollable quivers wracked through my body.

For a while, I could not say a word. Shaken by our narrow escape, I realised that we would have flown straight into the hill if Finn had not seen the trees above us. At our approach speed, the impact would have killed us all. Later, he revealed that he had only made out the dimly lit up shape of trees, by the light from the red rotating beacon, that flickered dimly off leaves ahead and above us. He said he "freaked out" when he realised what was about to happen.

Without saying a word, I flew overhead the LZ, which I orientated to the other lights in the area. When I picked up the direction the T faced, I aligned the pointer to this heading on the DI before turning downwind onto its reciprocal heading. After completing an approach and landing, while still not saying anything to Bob, who just sat in silence looking straight ahead, I took off again. We then headed back to base, and once we had climbed above our safety height, I handed the controls back to Bob for the return flight.

When I debriefed Bob, he agreed that flying helicopters was not his forte. I then told Finn that I was taking him for a few stiff brandies and broke all regulations when I took him into the bar at the officer's mess. Although inviting non-commissioned officers to the Officers' Mess was not permitted, I was so grateful for his vigilance that had saved our lives, that I ignored this regulation.

There is no doubt that if the two of them had been

on their own, despite Finn seeing the trees, with Bob's limited hours on helicopters, he would not have reacted as instinctively and decisively as I had.

After our near mishap and Bob not having his heart set on helicopter flying, I submitted a recommendation to terminate his conversion. He subsequently returned to No 4 Squadron, where he continued to do a good job flying the Provost, Trojan, and Lynx fixed-wing aircraft.

That was the first time I had looked the approaching prospect of imminent death in the eye.

15 AN AMERICAN CALLED JED

Jed was different to anyone we had ever met. But I suppose that was not difficult, considering the sheltered and conservative life we had all led in one of last bastions of the British Empire. The white Rhodesian population during the 1970s, was mainly of British descent. Prime Minister Ian Smith had declared UDI less than ten years before. Now the 300,000 whites were busy with what would later prove to be a futile and costly endeavour. They believed that they could hold back the aspirations of six million indigenous people and their ruthless leaders, who were willing to resort to any means to satisfy their hunger for power and wealth. To this end, these aspiring leaders managed to whip up nationalistic fervour.

The white population had done a fantastic job in less than a hundred years since the pioneer column's arrival. With only a few people and minimal resources, they had managed to hold off the Matabele and Shona uprisings. Since those early days, we had enjoyed many years of peace while building modern cities, with efficient infrastructures that rivalled countless first-world countries. The governing bodies ran the country in the well-established British pattern, and the BSA police epitomised the best in the world. The army and air force had also maintained close links with their British counterparts and incorporated the same regulations for the most part.

During the Second World War, Rhodesia's white population was so loyal to the Crown that almost every male volunteered to support Great Britain, and the authorities had to take decisive action to hold back a sufficient workforce to retain essential services.

During this period, the Rhodesians and New

Zealanders had established close bonds, as populations of both countries had a sizeable percentage of Scottish blood flowing through their veins, and they followed the same religious principles, values, and code of ethics.

Before Jed came onto the scene, Air Force pilots' worst potential fault was unreliability or slackness in fulfilling their commitments. Air Force officers worked hard and played hard. Now and then, they would be expected to attend a dining-in night and would need a valid excuse to miss one of these functions. At these dining in nights, they would all look dashing and smartly dressed in their formal mess kit attire and miniature medals on arrival. The first time I joined one of these dinners, I was horrified and bemused by their childish behaviour. During these events, the otherwise mature and professional officers acted like misbehaved children at boarding school. They would pelt one another with pieces of bread rolls and dollops of butter or try to get butter pats to stick on the ceiling. In retrospect, I shudder at what the smartly dressed waiters must have thought. Then, once the otherwise supposedly strict formal etiquette and formal requirements were over, the function would deteriorate even further, to more inappropriate behaviour for supposed officers and gentlemen. They would become utterly sozzled and choose teams to compete in the crazy and potentially dangerous games of *Bok Bok'*. In these games, the first team would bend, one behind the other, with their heads tucked under each other's crotches. One by one, individuals from the opposing team would then run up and leapfrog as near to the front of the bent-over team as possible. The aim was to overload some area of the bent-over team's backs without falling off. The losers would either be the leapfrogging team if one of them touched the ground, or the crouching team if they collapsed.

At one of these dining-in nights, I realised that I had consumed far too much alcohol. Recognising the onset of intoxication and knowing how much I would suffer the next day; I sidled up to S/Ldr Mike Saunders to inform him that I did not think I would be in any fit state to come to work the following day. This suggestion was a big mistake, as he duly castigated me in no uncertain terms and pointed out that an Air Force officer should drink according to the following day's commitments and needed to be disciplined enough to temper his intake of alcohol accordingly. I have only learnt more recently that I have an unusually low tolerance to alcohol and have suffered intolerable hangovers. But I never forgot that lecture. That night, when I went to bed, I had to keep one foot on the ground to stop feeling as if the bed was spinning around. The next morning, I could hardly sit up long enough to put on my socks and took over an hour to stagger gingerly to work. I did not manage to get to breakfast, and my head burst with each step.

As it was a Saturday, we came to work in civvies, in which we wore jacket, tie and hat, which we would doff instead of saluting. Our first stop was to the briefing room, where we assembled at 0800 hours, and which I just made on time, although still pale and shaky.

S/Ldr Don Brenchley, an ex-Royal Air Force navigator, gave us a precise time check, so we could synchronise our oversized Air Force issued watches. After the 'met' briefing was over, we strolled or cycled to our various squadrons, to nurse hangovers with morning coffee, which squadron pilots always drank with their feet up on the low circular Air Force coffee tables. They went on to recount the previous night's antics, but although most of the hilarity was at my expense, I did not participate in the discussions that day, but just slumped motionless in one of the crew room chairs.

The less affected pilots challenged one another to darts, as there was not normally a flying program on Saturday, and eventually, Mike felt I had learnt my lesson and sent me back to single quarters to sleep off the hangover.

The other part of Air Force life was for officers to attend Mess bar on Friday evenings, as a matter of duty. I think this tradition may have started in WWII, as a therapeutic way to cope with the loss of close friends and compatriots. Many other issues could also be resolved in the pub, as it allowed the opportunity to discuss grievances and let off steam. Unfortunately, the amount of alcohol one could put away without any after-effects was deemed to signify one's virility and masculinity. The general view was that only real men could imbibe copious amounts of alcohol with no ill effect. A favourite saying at that time was, "If you don't feel like drinking, *force* yourself." For many years, I assumed that I was not man enough or did not force myself sufficiently. But later in life I learnt that my lack of tolerance to alcohol was due to a medical syndrome.

The sad part of all this was that some young officers became alcohol-dependant after spending long hours in the pub, where it was the only time, they could feel socially accepted. Unfortunately, this environment was the breeding ground for those to become alcoholic. One such officer was a friend to everyone, and from those early days had a reputation for being able to consume up to 12 beers a night while on bush deployment, and still be fresh enough to fly his helicopter the following day. In his thirties, he continued to look like a teenager. Then, some years later, I bumped into him at Johannesburg Airport. I was travelling to the UK, and he was going to the USA. Accompanied by two black Zimbabwean Air Force wing commanders, he was delighted at the prospect that we might be travelling on the same flight and gleefully called me over. As we stood chatting, I was sad to notice that the years of alcohol abuse had finally caught up with him. In the end, he, and a few other well-known friends of mine, drank themselves into an early grave.

Despite a somewhat distorted approach to life, there was still what some might perceive, a questionable unwritten code of ethics. At the top of this list was, reliability, punctuality, being a team player, and never divulging another person's marital status while he was away on a night stop. The worst sin of all, was to tell someone else's wife what her husband had been up to while he had been away.

Then, Jed arrived on the scene.

We were told by the Squadron Commander that Jed was an ex-Vietnam, helicopter gunship pilot from the USA, which stirred up a bit of excitement when we learnt that he would be joining No 7 Squadron. So, we rallied around and clubbed in to buy and donate furniture and other household necessities. We also managed to secure one of the four New Sarum married quarters near the Officer's Mess for him and his young family.

From the moment we met him, Jed came across as a real extrovert and an unusually short character, with an Andy Capp belly. In uniform, his chest was adorned with four rows of brightly coloured medals. As a novelty, he thrived on being the centre of attention and enjoyed playing up to a captive audience, who were fascinated by his turn of phrase and American accent.

As the 'A' Flight Commander and Chief Flying Instructor on No 7 Squadron, I undertook his conversion onto the Alouette III. Two things immediately took me by surprise. The first was his aggressive and coarse language when I informed him that I would do the take-off, and then hand the helicopter controls to him once we were airborne. The second was his lack of flying ability.

On this first sortie, I took him on a familiarisation flight south of Prince Edward Dam, to our general helicopter flying area, in the Seki Reserve, were we used a whitewashed square, set up on the side of a *vlei*, to practice circuits, landings, and other various hovering exercises.

When I told Jed I would do the take-off, he went into a rage. He certainly took offence at the intimation that he was not capable enough to take off unassisted from the concrete helicopter pad. In shock and surprise, I submitted to his protestations. But fortunately, I had learnt from experience to keep my hands conveniently close to the flying controls, by resting them on my knees while instructing. Our helicopter was surrounded by other Alouettes parked on the pad. The moment we left the ground, we were out of control and careered around in the confined space, where we narrowly missed colliding into one of the other machines before I managed to snatch the flight controls from him. I was stunned by this unexpected lack of coordination from a theoretically seasoned, veteran, Vietnam pilot. However, I was more astounded by the tirade which followed my audacity at taking the controls from him.

Despite my state of shock, I managed to keep my cool, but I was continually taken aback by his total lack of competence over the next month, and wondered if this was the standard for US pilots in Vietnam. I was also amazed that he had even managed to survive an operation tour there.

After a week of instruction, he had improved enough to allow him to fly solo. I then decided to familiarise him with different parts of the country in a more operational environment. To achieve this, I organised a 'round the houses' trip to practice map-reading over diverse terrain. We spent the first week at FAF One at Wankie, before going to Kariba for one night, Umtali, and then Melsetter, where we could train mountain flying in the Chimanimani Mountains.

At Wankie, once again, his sparkling personality proved to be a hit with the junior territorial camp personnel. But some of the officers were shocked when they overheard his aggressive and colourful language during my pre-flight briefings, whenever he felt I was treating him like a novice.

I was uncertain how to deal with his unusual behaviour. But I persisted with the standard pedantic Air Force procedures I had learnt during my flight instructor training.

Apart from the abusive language which accompanied each pre-take-off briefing, a few extra surprises were in store for me. I first introduced Jed to 'confined area' operations and chose an open *vlei* area surrounded by trees. The first one I selected, was twice the length and width of a football field, and after his Vietnam experience, I expected would be purely academic. But he astonished me when he flew into one of his outbursts and demanded to know how I expected him to get in and out of such a small area.

I chose to ignore this flare-up, but the flight engineer and I exchanged bemused glances. I then continued to show him the safest technique in and out of confined areas, where I restricted his power to simulate operating at 'max all-up weight'. Throughout this exercise he complained about how small the clearance was, even though 50 Alouette helicopters flying in formation could have landed there at the same time.

In the pub that evening he continued to impress the youngsters on National Service call-up, which made him even more insufferable and self-important. It was also the first time I witnessed how much he fascinated the opposite sex, who were bored with the local talent.

The next day, I decided to practice some *'quick stops'* and autorotation exercises while flying low level. During the next pre-flight briefing, I spent some time warning Jed of the danger of striking the tail rotor on the ground in a high nose-up attitude. Once I finished my briefing, we headed off, just east of the runway to the Lukosi River, to provide stretches of some flat ground over the dry riverbed, and clear of obstacles for the planned exercise.

I first did an area survey, where I flew overhead to choose a straight section of the river, which was about 1,000 metres long, and clear of obstructions or overhead wires. When satisfied, I flew down between the treed river embankments to demonstrate some quick stops. Once again, Jed was impatient with my pedantic approach to such a simple exercise and intimated I was trying to "teach my grandmother to suck eggs". When it came to his turn to carry out the same exercise, he took so long to slow the aircraft down that we turned a bend in the river beyond our previously inspected area. When I told him that he was taking far too long and moving out of the planned area, he went into another one of his angry outbursts and assured me this was not a problem. But, in the middle of his protest, I caught sight of electricity cables stretching across the river just in front of us. Jed had not seen them, and I quickly assessed that if we tried to stop the helicopter by putting it into a high nose-up flare, we would strike the wires with the front of the rotor blades, or the riverbed with the tail rotor. Also, realising that we were too close to climb over them, I calmly told him I was taking over the controls and kept the nose attitude level while lowering the collective. In a flash, we shot beneath the overhanging wires with the wheels skimming the sandy riverbed.

Once clear of the wires, I climbed the aircraft to a few thousand feet with my knees turning to jelly at the realisation of what might have been. Still oblivious to the wires, Jed started to scream at me for rudely whipping the controls out of his hands. Without a word, I returned to Wankie, feeling I had suffered enough trauma and abuse for one day. But once on the ground, I immediately sat down to compose a report to the Squadron Commander, in which I told him that I had decided to complete my obligation and finish the scheduled training programme. Still, I did this with the proviso that if there was no substantial improvement in his attitude and performance, I had no choice but to throw in the towel and discontinue his training.

In due course, we went on to Kariba for one night and then to Umtali, where Jed managed to seduce a young woman at each place and would then boast about his conquests.

At Melsetter, he allured another local girl with whom he developed a more long-lasting relationship.

Despite having a formal sit-down discussion with Jed about his unacceptable behaviour and trying to impress on him that we were unaccustomed to such language and disrespect in the Rhodesian Air Force, he made no effort to change. I may just as well have been speaking a foreign language.

When I introduced him to the techniques used for confined landing areas in the Chimanimani Mountains, he again blew up into one of his fits of rage. Before this, I had taken many new helicopter students in and out of the same areas without a second thought and remained bewildered by his unwilling attitude to each exercise.

In planning for this mountain flying trip, I had phoned the Melsetter police station for permission to park the helicopter on their football field and requested a night-time guard. I also arranged for sufficient 200-litre drums of Jet A1 fuel to be delivered there. But when we arrived at the football field to start the next day's flying programme, we immediately got off to a bad start. While rolling one of the fuel drums to the helicopter, the flight engineer sliced the side of his finger on the seal covering the screw-on plug. He always looked a bit anaemic, and I am sure he had a few too many drinks the night before, as when he saw blood pouring from his finger, he fainted on the spot.

Once I administered some first aid, we departed to do some hoisting in the mountains, where we took Bob Halluch, the Chimanimani Hotel manager, an Air Force VR (Volunteer Reserve) to assist. On many occasions, we had previously winched Bob up and down in the strop at the end of the hoist cable to lower him onto inaccessible ledges. Then, after flying a circuit, we would reposition overhead the drop-off position, to bring him back into the cabin.

The strop that was connected to the end of the winch cable would normally be placed under Bob's armpits by the flight engineer, who would then electrically lower and raise him as the dummy. At the same time, the engineer would give the pilot a running commentary of the proceedings and provide instructions to help him hold a steady hover over the required spot.

Usually, once Bob was ready to be lowered, the engineer would move him out onto the port sidestep under the winch mechanism. Then when ready, Bob would spread out his arms and hang free of the step, before giving a thumbs-up signal. The flight engineer would also watch the tail rotor to keep it clear of obstructions during the whole procedure.

Each time I pointed out the ledge we would use during this sortie, Jed would subject me to the now-familiar outburst. To further exasperate me, the flight engineer, probably still suffering from the night before, also started to baulk and shake his head in reluctance when I asked him to lower Bob on the winch. Eventually, Bob also picked up negative vibes and showed resistance himself. This surprised me, as he had never hesitated before. Then, when he was supposed to release the leather handle at the door, and hang free of the step, Bob unexpectedly stood frozen at the doorstep. In the meantime, the engineer lowered the strop, which dropped from under Bob's arms to below his knees. The sight of him poised at the door with the strop hanging loose at the bottom of his legs, nearly freaked me out and gave me nightmare visions of his tumbling backwards out of the helicopter and plunging upside down onto one of the highest peaks in Rhodesia. Once again, I decided I could take no more of this travesty and called off the exercise.

After we left Melsetter and arrived back at the Squadron, I reported the whole sad story to the Squadron Commander, S/Ldr Eddie Wilkinson. But he, in his dubious wisdom, decided to give Jed another chance and suggested that the other Squadron instructor, Flt Lt John Annan, should take over Jed's training.

To John's credit, he was either more tolerant than me, or managed to control Jed's temper better than I could. Regardless, John completed his conversion to a questionably low standard for an operational pilot, and the Squadron put him on operations. Nevertheless, Jed never stopped being a liability and managed to upset numerous people along the way. On one deployment to Mushumbi Pools, he was the only helicopter pilot attached to an army unit. During a poker game, with his continual chirping, he upset an army officer to such an extent that he was sent flying off his chair with an angrily swung punch.

When the next ex-Vietnam pilot, Bill McQuade arrived, I expected the same low standard. But was amazed by him when he competently managed to lift the helicopter off the ground into a rock-steady hover, right from the start. The difference has subsequently led me to suspect that Jed could not have been a qualified helicopter pilot at all. At best, he might have been a gunner in one of the Huey Cobra gunships. However, I will never know how he produced all the relevant documentation and a pilot's logbook.

16 MUSHUMBI POOLS

I met Jim Barker during my time on helicopter operations in the bush. He was a charismatic character from the Karoi area, northwest of Salisbury, and was a successful farmer who owned a light aircraft which he made available to the Police Reserve Air Wing.

Before we met, I had read the story about his wife Judy, and her two small children, who had come face to face with a group of terrorists at their farm store. Their young son had a BB gun (toy replica pellet gun) which he threatened to use against them. Fortunately, the family was able to depart unscathed.

Sometime later, Jim invited No 7 Squadron pilots to a party at his farmstead and offered to collect their wives in his aircraft, take them to his farm for the party, and return them the next day.

Jim's offer fitted in well with a request to deploy a helicopter to Mushumbi Pools on the Hunyani River to support an army unit there. So, we planned the party for the previous day. Fortunately, it was not far off the direct route from New Sarum to the army unit's base at the bottom of the Zambezi escarpment. As it was just to the north of Jim's farm, we could justify using two helicopters by organising a navigation exercise called a 'flag-ex,' and finish at his farm for the night. Extra pilots flew as passengers in the second helicopter, which returned to base the next day, while I went on to Mushumbi Pools.

Rhodesian farmers were renowned for their hospitality, and Jim and Judy were no exception. Farmers and their families endured incredible hardships to continue productive farming and displayed great courage to live with the ever-present threat of land mines, ambushes, and night attacks on their homesteads, although, most often, they managed to keep the attackers at a distance from the farmhouses. The government contributed to the erection of security fences, and bright security lights around the homes, and the perimeter fence barriers afforded a measure of protection. Each farm also had an '*AgricAlert*' system installed, which provided a direct radio link to the nearest police station, and every morning and evening the local police station would conduct a roll call to check each farmer in the district. In an attack, the farmers could immediately call the police, who would call for a reaction force to dispatch to the area.

Fortunately, the terrorists seldom displayed any real courage and preferred to strike soft targets where there was no possibility of a quick security force reaction. Consequently, the enemy would subject farmers' homesteads to hit-and-run attacks.

That night at Jim's farm, the wives enjoyed an opportunity to enjoy a party with their husbands. The adventure of this visit to the Barker farm, gave them a break from their monotony at home. By the time the helicopters arrived, they were already in a festive mood, and Judy also enjoyed having the company of the visiting wives. Lunch was a sumptuous braai around their sparkling swimming pool, while we enjoyed dancing to the latest pop songs on their spacious veranda.

Judy made a wonderful meal for dinner, starting with a mock crab cocktail and seafood dressing. The food was delicious, but I am unfortunately allergic to seafood, which affected me badly the following day when I had to make an early departure for the Mushumbi Pools base camp. During the flight, I had to land the helicopter in an undeveloped area, with the onset of diarrhoea, and by the time I reached the camp, I was ill. To add to my discomfort, the temperature in the shade was around 45 degrees Celsius, with no wind to alleviate the oppressive heat.

Over the next few hours, I had to continually rush to the toilet, which was a 'long-drop,' built on the top of an anthill and surrounded by hessian strung around cut poles. It was out in the open, with no roof to shield the occupant from the blazing hot sun and sitting on the wooden box perched over a foul-smelling hole, I also had to endure the mean-looking flies that continually buzzed in and out of this pit.

The loss of fluid from the diarrhoea dehydrated me and sitting on the 'long drop' toilet with sweat pouring from my forehead and armpits, further aggravated the situation. The army unit provided respite from this heat by filling a small free-standing canvas pool made available to anyone wanting to cool off, but I found it quite distasteful when I tried it. Apart from doing little to help in the open, the scorching sun made it feel more like getting into a hot and dirty bath.

I was grateful when the medic gave me pink Kaolin medicine and heat fatigue tablets, which went a long way to combat the diarrhoea. Still, by that afternoon, I could not take the oppressive heat any longer. So, I told my engineer that we were going on a flight.

We removed all the doors, and as we were the sole occupants, the aircraft was exceptionally light. I then climbed the helicopter up to 10,000 feet, where it eventually felt like being in refreshing air-conditioning. At that altitude, the outside air temperature had dropped right down to 18 degrees Celsius. Regulations did not permit us to fly above this height without using oxygen equipment and a mask. But it was an enormous relief to be able to fly around for over 30 minutes and escape the relentless heat at Mushumbi Pools.

The Army units often thought that pilots were lounge lizards, and I gave them sufficient reason to feel that way. But once more, I was able to put the helicopter's versatility to terrific use!

17 FIRE FORCE ARRIVES AT BUFFALO RANGE

The K-Car with 20mm cannon

Towards the end of 1975, I had served nearly six years on No 7 Squadron. The Rhodesian bush war had been raging over the last three years and was spreading out of control. It had now reached as far as the sugar cane and citrus growing area of the Lowveld. However, by 1974, it seemed that the security forces had dealt a severe blow to ZANLA's infiltration and their attempt to establish themselves in the northeast of the country. By that stage, they were close to wrapping up Op Hurricane.

Then a *coup d'état* took place in Portugal on 25 April 1974, and this overthrew the Prime Minister, Marcello Caetano. On 27 July, the new President, General António de Spínola, recognised independence for the overseas provinces of Mozambique, Angola, and Portuguese Guinea. He immediately put them on the road to self-determination and ordered the Rhodesian Government to cease all cross-border operations. Shortly after this, the US Secretary of State, Henry Kissinger, turned the screws on the South African Prime Minister, John Vorster, who obligingly withdrew all military personnel and financial assistance to Rhodesia, when all parties supposedly signed a ceasefire. But unfortunately, this merely gave the insurgents an opportunity to lick their wounds and re-group for a concerted and more coordinated push. Their next return was with a vengeance, which included harsh punitive measures being taken against any villagers deemed to have been 'sell-outs.'

During the previous ten years, with South Africa's support, the country had been able withstand stiff economic sanctions and Aerospatiale, the French aircraft manufacturing company, had continued to supply us with Alouette helicopters and sufficient spares to keep them flying. Further support and assistance had also been provided by the South African government, who sent pilots and Army personnel who operated in the guise of the South African Police (SAPs). While there, they wore distinctive camouflage kit, more suited to the desert than Rhodesia's indigenous woodlands. The South Africans also contributed additional Alouette III helicopters and a few light transport aircraft to the war effort. However, as part of the ceasefire agreement motivated by Henry Kissinger, the South African government withdrew all their military personnel. But, as a gesture of support, they left their helicopters behind.

When the war began at the end of 1972, the Rhodesian Air Force started anti-terrorist operations with only six Alouette III helicopters. But after six years, including the helicopters left behind by the South Africans, the Alouettes at our disposal had increased to forty. This meant, that to fill the gap left by the withdrawn South African helicopter pilots, we had to bring back previous operational Alouette pilots from other squadrons and train up new recruits.

By then, as the A flight commander, I was responsible

to rotate and deploy forty pilots in and out of the field. Then, when pilots were back from operations, I also had to ensure they were adequately trained and competent in all other aspects of helicopter duties like instrument and mountain flying, to name a few. I also had to provide sufficient crews on standby duty for unexpected tasks that would crop up at all hours of day and night.

I was part of a group of pilots who had to adapt quickly and develop tactics and modus operandi to counter the growing insurgency problem. Having had this early experience, I was better equipped to prepare new students during their conversions and operational training. By the end of the 70s, the growing need for more helicopter pilots, resulted in students being posted to the Squadron directly after receiving their wings.

After joining the Squadron, students were thrown straight into an intensive conversion onto the helicopter, during which they were introduced to carrying out simulated running battles among the granite outcrops in the Domboshawa area north of Salisbury. This challenging area improved their map-reading skills and introduced them to our evolving operational techniques. After only two-and-a-half months, the Squadron sent these new pilots to support army and police units in their ongoing counterinsurgency operations.

In the early days, students benefited by the instructors accompanying them on a 'bush orientation' period to ease them gently into the operational environment. But as ever-increasing numbers of insurgents entered the country, this luxury fell away. The changed situation plunged the raw pilots straight into fierce running battles from their first day in the field of operations. That they survived says a lot for the training syllabus, which we continually adapted to meet the escalating conditions.

Before the South Africans pulled out, their helicopter pilots were stationed at the more 'plum' holiday resort locations like Kariba and Wankie. But as the attacks increased, they joined up with our pilots in the operational areas. Nevertheless, their helicopters remained painted in their unique camouflage. which supposedly depicted them as being part of the SAP (South African Police).

When we first introduced these pilots to active operations, we preplanned to drop off troops at preselected drop-off points in an area suspected of holding some enemy forces. As the briefing progressed the day before a first-light encirclement of a group of thatched huts in a rural village, I looked across at Capt Flip Van Zyl, a charismatic South African pilot, with a long drooping moustache. While the army commander allocated where Flip would have to drop off his stick, he realised in horror that it would be right in front of a hut reported to accommodate a group of terrorists each night. At the time, he was busy smoking a cigarette, but was so taken aback by this information that his eyes shot wide open, and he gave an involuntary inward gasp. This sharp intake of breath caused him to suck so hard on his cigarette, that it brightened the glowing tip which rushed noticeably towards his pursed lips.

Flip had already amused us the day he arrived at Mount Darwin to start his first tour of duty. The ops room sent him on a task with no time to settle down or familiarise himself with the area. Then, on the way back, the relay station called him to divert him to a new position. They relayed this message by using 'shackle,' which was a simple code that the security forces changed daily.

The 'shackle' principle was a code used to substitute the numerals 0 to 9, with letters of the alphabet, a system the security forces introduced to prevent the enemy from hearing grid references passed over the radio. Every day, pilots would get the 'shackle' of the day on a strip of paper and when they, and other security forces, gave the grid reference in letters, they would then be decoded into numbers to locate the correct position on a map. This action added to an already heavy load as they would have to accomplish all this while still flying the aircraft before leaving their present location.

Flip was called by the relay station over the radio, "Yellow 3, could you pick up a stick of *'troopies'* from, shackle 'Uniform, Sierra, Alpha, Charlie, Hotel, Oscar, Sierra, Papa?'" Already fed up with his first day, he retorted with some irritation in a broad Afrikaans accent, "Don't come with that *Tachnical* stuff! *Yus* give me the *nummers*!"

The 'fire force' concept was a tactic developed over time. It proved more successful, as it provided more flexibility and made better use of our limited resources. It was also a way for the army commander to efficiently utilise the forces at his disposal, when they were brought into situations where someone had become aware of some enemy presence, or there was an ongoing 'contact'. Here the helicopter pilots could uplift troops but delay dropping them off at any preselected positions.

The Army's first troops employed in Fire Force action were the crack troops from RLI (Rhodesian Light Infantry). They would be on standby at the base, already kitted out in khaki shorts and black canvas hockey boots.

All that was necessary was for them to pick up their radios and weapons. However, before we fully developed this idea, we relied on intelligence reports of the enemy position from OPs (observation points), airborne reconnaissance, and intelligence gained from the Selous Scouts, acting as pseudo-terrorists.

Before the security forces developed the fire force concept, the planners would allocate each pilot a pre-designated place to land his helicopter. They would give the pilots the exact spot to drop off their sticks during the ops briefing before take-off. They planned for the airborne contingent to carry out many of these initial deployments to suspected locations during dawn raids. However, this was not very successful as it committed the helicopter 'stick' to a specific location and a restricted area after the helicopters had deployed them. In most cases, the terrorists had fled or 'bomb-shelled' in all directions at the sound of an approaching helicopter formation. To overcome this, someone came up with the bright idea of using the Trojan aircraft's noise to shroud the distinctive high pitch engine's whine and clattering helicopter blades.

The AL-60F-5 Trojan aircraft, built under licence by Aermacchi and designed to carry passengers, was not endowed with anywhere near perfect aerodynamics. Its piston engine was exceptionally noisy and grossly underpowered when operating at the height and temperatures found in Rhodesia. Apart from transporting personnel around the operational area, security forces used the Trojan extensively as spotter aircraft. After some time, the Air Force also fitted it with guns, rocket pods, and other weapons to assist in the ground attack role.

The pilots on Four Squadron did a remarkable job with these aircraft despite their limitations. But sadly, pilots on other squadrons made derisive jokes about the Trojan's performance and noise level. One was, "The only reason a Trojan can get airborne is because of the curvature of the Earth"; or "The engine noise is made by the five pistons clapping the one doing all the work!"

Trojan pilots became highly skilful in airborne reconnaissance. They could recognise distinctive tell-tale networks of footpaths converging under thick canopies of trees that concealed terrorist base camps. Some pilots became so talented that a few, like Cocky Beneke and Peter Petter-Bowyer, could recognise individual camouflaged sleeping positions called 'Bashas,' which were crudely constructed frameworks saplings under thatched grass. I think that Cocky might have had a slight colour vision defect, which might explain how he could discern camouflaged objects so easily.

Trojan aircraft were also used extensively as spotter aircraft and airborne relay stations (Telstar) to pass messages between base stations and troops on the ground. After a while, insurgents became blasé about these aircraft that continually circled overhead without posing any threat. This fact prompted the decision to use the familiar noise of an overflying Trojan to conceal the approach of helicopters bringing in the fire force. Inevitably, the enemy soon cottoned on to this tactic and became suspicious whenever a Trojan flew overhead their position!

The Air Force planners decided to fit a 20mm canon onto a few Alouettes, which enhanced the fire force's effectiveness. After the helicopters had deployed their sticks, the one with the mounted cannon would have the commanding officer (Sunray), who could keep a watchful eye from overhead and give immediate backup firepower for the much feared and respected RLI troops.

The RLI Battalion was initially formed by recruiting youngsters who were dropouts of society. But their thorough training moulded them into an elite and close-knit family of super-efficient young soldiers. They were courageous and very successful in battle in the fire force context. In time, they became an efficient and proud unit, and developed a jargon incomprehensible to most outsiders.

A formation of helicopters that took them was led by the 'K-Car' with its mounted, 20 mm cannon, with a forward-facing front seat for the ground force commander (Sunray). At the same time, the gunner operated the weapon from the rear of the cabin. The security forces called the helicopters that carried up to six troops the 'G-Cars.' The Air Force fitted them with .762mm MAG guns mounted on the left side, which the helicopter engineers operated.

Sometimes, the fire force followed an air-to-ground attack by Four Squadron's fixed-wing aircraft. But when the target was an occupied base camp that housed a large group of insurgents, the JOC would utilise Canberra bombers from No 5 Squadron and Hunters from No 1 Squadron. These would saturate the enemy position with some locally developed anti-personnel weapons with devastating effect before the helicopters arrived.

Later, the Army and Air Force personnel decided that the G-cars should delay deploying their troops after arriving at the scene. They would then circle the area until the airborne commander and the K-Car formation leader had carefully assessed the situation on the ground. Apart from providing an extra lookout, this development offered better flexibility.

On arriving at the location, the formation of helicopters immediately entered a left-hand orbit, making it possible to bring the MAG and 20mm cannons mounted on the port side into play. The helicopter pilot and everyone on board also provided extra 'eyes' from above, and flight engineers could fire the guns if anyone saw the enemy.

The K-Car formation leader and the onboard army Sunray would make decisions together. The K-Car, the formation leader, would then direct each G-Car to the position to drop off his 'stick.'

Once the G-cars had dropped off their loads, they would either orbit the area to assist the ground troops further or return for backup forces. Depending on how the contact developed, the K-car could also direct G-Cars to drop the onboard troops to more appropriate locations. The next load of soldiers was either uplifted from the base camp, or from a preselected staging point, where extra drums of Jet A-1 fuel had been brought.

During a call-out, the K-Car pilot and Sunray could

spend many hours directing troops, providing backup fire when needed, or firing at ad hoc targets seen from above. Each 20mm shell was as effective as a hand grenade, which would explode on impact, with devastating effects, up to three metres. Despite the excitement and confusion of battle, many gunners became so proficient that they often scored direct hits from the helicopters, generally orbiting at 800 feet AGL. Depending on the terrain, the K-Car sometimes directed G-Cars to carry out dummy drops to make fleeing terrorists think they were deploying troops in ambush positions. This practice would often deflect the direction of escape and channel them towards newly established 'stops' (ambush positions) selected by the K-car airborne commanders.

As the K-Car continued to orbit overhead to direct 'call signs' on the ground, it remained exposed over long periods and would sometimes come under heavy ground fire. On one contact, I was aware of the continual 'ticking' sound from rounds narrowly missing the aircraft. The accompanying 'sunray' was a handsome young army major, who a well-known journalist had recently interviewed. Stricken by his good looks, she poetically described him as "six-foot-two and eyes-of-blue" in the magazine article. He was more comfortable operating on the ground, where he felt less exposed behind a blade of grass than now, as we came under continual enemy fire. While I orbited the area and directed the G-car pilots where to drop their sticks, he became disorientated. He tucked his long legs into the armour-plated seat to present as little as possible to the enemy gunfire. In a state of sheer panic, he started to scream at me, "Let's get the hell out of here!" His shouting made it difficult for me to concentrate on the task at hand, or give directions to the G-Car pilots, so I yanked out his microphone's headset lead from its overhead socket to shut him up.

Over time, helicopter pilots were involved in many contacts and saw more action than most of the individual army fire force commanders. Through this exposure, they became proficient at implementing effective tactics or deciding where to set up stop positions as the battles progressed.

Sometimes these contacts would rage on for the whole day, so I quickly developed the habit of storing rations under my seat. This included ration pack biscuits (cynically called "dog biscuits"), tins of bully beef, and fruit salad in readiness for when there was no chance to have breakfast or lunch. Apart from this, I periodically nibbled on one of these dry biscuits as we flew and was then able to have something to eat each time we landed to refuel.

Over time, as the army became thinly spread out on the ground, more and more units would create their local fire force. Eventually, Territorial Army units on call-up would also operate in the fire force roll. Later, the Army had to withdraw the SAS from their primary role of parachuting deep into neighbouring countries to sabotage bridges and other targets or generally harass the enemy. The SAS would later deploy into local battle areas by parachuting from Dakotas. This proved so successful that the RLI and other units followed suit, after being given parachute training. The DC3 aircraft continued to drop these troops from as low as 800 feet AGL, both day and night.

After many different callouts over the years, pilots could immediately recognise unusual behaviour patterns while circling overhead a contact area at 800 feet AGL. One day, during the firefight I was busy with a conflict around an African village complex, when a lone individual in a blue shirt caught my eye. Even from above, there was something suspicious about how he kept looking up nervously while trying to get away as quickly as possible without attracting attention. His behaviour did not fit the scenario of the locals who were scattering in all directions.

As he half-walked and half-ran, I had the impression he was trying too hard to look like an innocent villager, who was not trying to escape. At that stage, although the terrorists were committing horrendous atrocities against anyone suspected of being a "sell-out", there were still strict guidelines as to when one could open fire. The regulations only permitted you to open fire on someone if you saw him firing a weapon or carrying one with intent, and anyone firing at an innocent person would have to face a court-martial.

Nevertheless, I was confident that this person was no ordinary villager and instructed my gunner to open fire on him. Of course, this sent him straight into a run as he tried to evade the shells striking all around him, until he was stopped dead in his tracks. I then asked one of the 'call signs', who had been previously dropped into the area, to check the body. Luckily, I had not been mistaken, as they discovered a weapon and webbing with rounds of ammunition concealed under his shirt. When the JOC debriefed me that evening, the army commander found it inconceivable that I could have been sure enough to commit myself from 800 feet above the ground.

The enemy had shot down a few helicopters and fixed-wing aircraft involved in combat missions from low altitudes. Most of the pilots ranged between the ages of mid-twenties to thirty. Despite the risk, they still got an adrenaline rush from running battles. It became more like a sporting activity, with close synchronisation, trust, and understanding between the helicopter crews. Morale was high, and combatants established a close bond. The same thing seems to develop between top sportspeople in successful teams and others during times of war. All of which could be the shared common purpose and elements of risk.

In the early days, the biggest threat to airborne operations proved to be the dreaded Strela (SAM 7), a Russian hand-operated anti-aircraft missile. One of these brought down Chris Weinman, one of the pilots on No 4 Squadron.

He had been doing airborne reconnaissance from a Trojan while flying over a thickly wooded mountainous region in Mozambique, just north of the Zambezi River and opposite the Mague airfield. At the time, no one knew what had happened to him when he failed to return to base and the next day, we all took part in an intensive search of the area where he had been operating. The Air Force deployed several helicopters and light aircraft to look for places where he might have come down. Not sure of the exact location, we spent many hours deep in enemy territory, flying low over the thickly wooded and sparsely populated mountainsides and gorges. Here we were exposed to considerable risk when we flew at 45 knots to get a closer look at gullies and rugged terrain. When we saw a Strella hitting another Trojan taking part in the search, flown by a fresh young pilot, Air Sub Lt R Wilson, with Flt Sg Rob Andrews, we saw where it crashed. With a concentrated focus on the same area, the search team eventually found Chris's aircraft, which the enemy had carefully covered with cut foliage branches. Sadly, neither Chris nor the crew of the second aircraft survived.

At another time, we were operating out of Mague on the southern bank of the Zambezi River. While I was busy flying troops to an area north of the river, I spotted a semi-decayed elephant carcass lying in an open marsh in a 'hot' area. I noted the spot and returned to the airstrip for my last load. While the troops settled into the helicopter, I sent the engineer to fetch a *panga* (machete). Once we had dropped off our stick, I returned to the rotting carcass. The engineer got out to extract the tusks. But after struggling for an uncomfortably long time, I realised he would need assistance. So, I shut down the rotor blades and left the engine running to start quicker if we had to make a quick getaway. I hurriedly went to give him a hand and we finally managed to remove the tusks, with the stench of the half-decayed nerves turning my stomach. At the end of the two-week stint in the bush, we took the tusks back to Salisbury in black plastic bags. As I wondered how we could sell them and split the profit, I recalled that my widowed aunt's husband had been a big game hunter in the Zambezi Valley in Southern Rhodesia at the turn of the century, and she still had several unregistered tusks from that period. So, I contacted her to ask her if she would like me to sell them for her. She liked the idea and gave me some photographs and newspaper articles for the Game Department. These verified how she had come to possess so many tusks, which her husband had never registered. The department official was understanding and sympathetic and agreed to record and stamp the tusks. Fortunately, he was not at all suspicious of the recently acquired pair that I included, and I managed to sell the whole lot. My aunt was so grateful that she paid me a tip for selling her haul of unwanted tusks.

As the terrorist incursions spread further into the developed, agricultural areas, I held the white farmers and their wives in the highest esteem. They were always vulnerable to enemy attacks and land mines planted on their dirt roads. Nevertheless, they carried on with remarkable courage and continued their farming activities under risky conditions.

One day, I was with a few pilots who had positioned our helicopters in the eastern mountains near the Mozambique border for the night on a farmer's lawn, overlooking Cashel Valley. Here, we spent the night in the farmhouse, which had a clear view of the valley from the west-facing veranda, where the garden's edge dropped precipitously for 2000 feet. Air Force HQ had sent us to this farm to carry out a cross-border operation into Mozambique the next day. The cook took great care of us as the farmer and his family had moved out, after detonating a land mine outside their front gate.

As we sat around after supper, we heard a farmer's wife call the local police station on the AgricAlert radio. She calmly informed them that terrorists were attacking their homestead with mortars, RPG 7 rockets, and gunfire. After a short while, a Lynx aircraft from Grand Reef airfield, west of Umtali, arrived overhead and fired off rocket flares. These slowly descended on small parachutes and lit up the whole area. While we sat on the veranda, it was perfect view to see the flares lighting up in the valley below. Unfortunately, we were helpless as we were in no position to help. However, this young wife inspired me. She continued to give a running commentary on the proceedings as things developed. Sounding completely calm throughout, she could just as well have been discussing a chicken recipe with one of her friends as she continued, "An RPG 7 has just hit the top of the roof"; "My husband is at the front window returning fire".

We met many of these brave farmers. Some, of Scottish descent, were entirely ill-prepared for the tropics' farming activities, with ruddy skins, blond hair, and blue eyes. While the farmers supported units operating around the country, their wives tried to keep things going. Older men, named 'Bright Lights' were then provided as extra protection to the wives inside the security-fenced gardens. The irony of it all was that while the farmers took part and contributed their time towards the war effort, their brave young wives were supposedly kept safe by these older men, who were from urban areas in the country.

Most farmers were 'salt of the earth', and helicopter pilots befriended many. They would meet at the local sports clubs or get invited to their farms for tea or lunch. At other times we met them when we had to land on their farms. The farmers and their wives would always welcome us with gracious hospitality on every occasion.

Towards the end of my time on No 7 Squadron, security forces opened a new operational area in the Lowveld. I was the K-Car commander on the first Fire Force to arrive at the base there. The JOC established their headquarters at the Buffalo Range Airport, near the town of Chiredzi.

Within the first week, the local district Commissioner gave a talk to all of us in one of the larger tents there. His knowledge of the African culture and its traditions was vast. I learnt a lot about their courtesy through this talk, which made me realise how crass my bad manners must have appeared to our African brethren I had been in communication with before. He highlighted that they never immediately broached the main subject when they first started any discussion. They would always go through the greeting formalities and inquire about all family members. When one would have thought the conversation had finished, the one coming to discuss a matter, would introduce it, as an afterthought. They would also not enter a villager's private enclosed area without calling, *"Go, Go"* (Knock, knock). Since then, I have always asked people 'how they were' when I meet them.

I already knew how hospitable Rhodesian farmers were, but I would never have believed that anyone could overshadow them. However, the sugar and citrus-growing farmers of the Lowveld managed to accomplish this. The wives immediately set up a daily hot dog and hamburger stand on the airstrip, which they happily kept going the whole time we were there.

One day, Jean Claude, a French sugar farmer and a bachelor from Mauritius, invited some of us to a party at his house. While we could not buy wine and whiskey in the shops, Jean Claude somehow proffered an unbelievable supply of good South African Drosdyhof red wine and the unique triangular bottles of Dimple Haig whiskey. Harold Griffiths (Griff) and I headed off to the party in a Land Rover lent to us by John Digby, the Air Force representative at the JOC. Although Griff was my Squadron Commander, I was the appointed K-Car commander for that deployment. It was an unusual experience at the party to have such an abundance and selection of drinks, and I had such a good time that I lost track of how much alcohol I had consumed. I only realised just how intoxicated I was, on our way home that night. The confusing maze of roads between the tall sugar cane soon disorientated us. I offered to climb onto the Land Rover's bonnet and look for a glow of lights from Chiredzi. But as I tried to stand up on the bonnet, I fell backwards, straight into the spare wheel.

From previous experience, I knew how much I would suffer the next day. So, I quickly offered Griff my flights on the K-Car, which the JOC had tasked for an early departure. Fortunately, he was happy to accept this offer. I then took a few Panados to forestall the inevitable hangover, but I still had trouble closing my eyes without the tent spinning around. The other helicopters had departed long before I woke up the following day. I had overslept, and it was already late. With the sun blazing in broad daylight, this spectacle of a sick pilot amused everyone walking past my tent. They knew how hot it would be, with the sun baking directly onto the canvas roof. Thankfully, a sympathetic medic came to my tent flap with two large green tablets, saying he thought I might need them.

It took another hour before I could pull myself out of bed, and with just a towel draped around me and in full view of the hamburger and hot dog ladies, I moved delicately to the ablution block for a shower. To start off, I stood under a jet of water. But when this became too much effort, I sat on the shower floor. When this effort also made me feel worse and became too much for me, I lay flat on the floor for nearly an hour while the water drummed down from above.

By the time I had recovered enough to make it to the mess tent for lunch, Griff and the other pilots returned. As I sat there, delicately nursing an awful hangover, I looked forward to a good sleep that afternoon. John Digby, who was fully aware of my fragile state, sat at the table when Griff arrived. With some sarcasm, Griff chirped, "I suppose you had a lie-in?" To this, John replied with tongue in cheek, "No, George went for an early morning jog around the airfield!"

Hearing this, Griff retorted in bemusement, "You stupid idiot!" This remark tickled John, but he managed to keep our secret. What made him roar with laughter was when Griff asked if I would like to do the afternoon flight, and I quickly responded, "No thanks, I don't need the flying hours!"

It takes some people time to realise that they have a problem. Sadly, I am one of them. It would be another ten years before I was diagnosed with Gilbert Syndrome, which explained that one of the effects is intolerance to alcohol and why I have suffered so many debilitating hangovers.

18 MOUNTAIN FLYING TRAINING

Alouette Flying in the Chimanimani Mountains

When the Air Force posted me to the Helicopter Squadron, PB (Peter Petter-Bowyer) and Hugh Slatter had drawn up a well-established training syllabus. They had preceded Griff and me as the Squadron's helicopter instructors and incorporated jollies into the syllabus. In Air Force jargon, jollies were trips away from home, with ostensible training value. However, to cynics, these could have appeared more of a break from mundane Squadron commitments and just built-in opportunities to enjoy the luxury of hotel life at Air Force expense.

My helicopter instructor, Harold Griffiths, was thoroughly brainwashed by these forerunners and continued to promote the concept wholeheartedly. His approach to the job at hand made my conversion to helicopters an absolute pleasure. But I must admit that I knew I was onto a good thing! By becoming a helicopter instructor, I could appreciate the benefits flowing from this practice. I also employed the system in navigation exercises across large portions of the country, which always culminated with training in the Chimanimani Mountains.

These trips, which would last up to a week at a time, would occasionally stick in the throats of other Squadron pilots, as before the commencement of the bush war, all they could anticipate were two-week stints at the Rushinga or Kanyemba army bases. The attached South African pilots had robbed them of the 'plum' detachments to tourist locations like Kariba or Wankie. I suppose I could have felt more guilty, but someone would always need to churn out professionally trained pilots and prepare them for the unexpected challenges ahead. However, I thought that the perks and bonuses emanating from these 'jollies' were well-earned rewards for helicopter instructors' demanding rigours and responsibilities.

The trips to the Chimanimani Mountains in the Eastern Highlands of Rhodesia required organising hotel bookings and planning for drums of Jet A1 fuel to be positioned at various police stations around the country. The time away included stops at Wankie, Kariba, and Umtali before arriving at Melsetter, where we spent about four days at the Chimanimani Arms Hotel.

I enjoyed going to this location twenty times over the six years on Seven Squadron. Over that time, I became good friends with the hotel manager, Bob Halluch, his wife Linda, the police member-in-charge, Tiff Thomson, Melody, and Bob Bailey. Bob, and his wife Loraine ran the Outward-Bound

School at the foot of the steeply rising mountain range. They and the Outward-Bound instructors referred to us as "noise pollution." But they were still happy to come up for a ride in our 'noisy' machines in their capacity as police reservists.

I loved the damp misty mountain atmosphere of Melsetter, with forests of wattle and pine trees. Pork Pie Hill's dome overlooked the Chimanimani Arms Hotel entrance, with pink and purple flowering Azalea, Rhododendron, and Camelia bushes filling the garden. A large stone hearth with a roaring fire welcomed guests into the foyer area. Further on to the right was an authentic English pub. Old pictures hung on the stone walls above comfortable brown leather chairs and a pervading smell of burnt charcoal and being attended to by broadly smiling waiters.

The whole place looked different to anywhere else in Rhodesia and was more akin to the Scottish Highlands.

After settling into the hotel and whenever possible, we would make an early morning start. The first task was for two helicopters to carry underslung drums of fuel in cargo nets to The Sea of Grass, high up in a shallow valley, surrounded by looming majestic rocky peaks with names like Giants Castle and Turret Towers. These stately pinnacles overlooked crystal-clear streams, teeming with trout and meandering through the valley below.

By the time we returned to the hotel, the exertion and fresh mountain air always worked up an appetite, which added to the enjoyment of a hot English breakfasts cooked on a wood stove. We relished tucking into the fried eggs and mixed grill, which invariably included minute steak, boerewors sausages, fried tomatoes, and sauté potatoes. This was then followed up with steaming cups of hot coffee or tea, and marmalade on toast. In the meantime, the kitchen staff prepared picnic baskets with fresh ham, cheese, and tomato sandwiches, newly baked scones, fresh fruit, and thermos flasks of hot beverages. These we would take with us when we flew the helicopters back to the mountains. There, after a few hours of training, we would enjoy a picnic lunch at some breathtaking scenic spot.

One of the most reckless and scary traditions we continued to do, was to winch unsuspecting students onto a mountain ledge. Then, instead of bringing them directly back into the helicopter, we would winch them back up a short way. They would then be taken over the rocky crags, while dangling about ten metres below the aircraft. At the time, we thought it was a bit of fun, but we would have had some explaining to do in the event of a mishap.

The weather in this region is very unpredictable and does not follow the annual rainfall patterns experienced by the rest of the country. Because of this, we inevitably lost at least one and a half days while we waited for the low cloud or fog to lift. We occasionally sneaked in under gaps below the cloud and saddles in the mountain range. By doing this, we would be ready to fit in a few sorties as soon as the weather improved. Although in humid conditions this could sometimes prove risky because as soon as the rotor blades started winding up, it could cause the fog to thicken up so fast that visual contact with the ground would rapidly disappear.

When making an approach to land on a mountain ledge, up-and-down draughts were also unpredictable, and pilots could quickly end up facing potential tragedy. With this in mind, we would never attempt to land on a recessed area amongst rocks or lower someone down by winch without first doing a power check. This was done by flying parallel to the cliff face at 35 knots and the same height as the LZ. Then, as we passed abeam it, we would take note of the torque required to maintain level flight. Using the onboard computer on the torque gauge, we could then calculate the necessary power to hover out of ground-effect. By comparing this to the maximum power permitted at the current 'density altitude', we could determine if it was safe to continue. It was necessary to go through this exhaustive routine at each selected location, especially as the most benign-looking spot could sometimes catch you unawares and prove far too dangerous to attempt an approach.

Once we saw a possible landing spot, we would fly past to inspect it, parallel to the mountainside. By doing this, we could ensure a quick and safe escape route in an emergency. At the same time, we would note our heading and altitude before flying the circuit away from the mountain. Ideally, the downwind leg would parallel the cliff face on the reciprocal heading.

When flying in the mountains, visual reference alone, made it impossible to maintain the same level as the LZ. Particularly as one becomes so easily deceived by the sloping lines of strata and other factors. When opposite the LZ at the predetermined circuit height, one often feels above or below it. These easily contribute to optical illusions that give false impressions of the natural horizon, making it essential to maintain the correct circuit height by referencing the altimeter alone. I have flown with students, and even a visiting Aerospatiale test pilot, who lost over two thousand feet while trying to keep level by visual reference.

If everything went well, we would delay the turn towards the landing site until the very last moment. The final approach to the LZ would be flown through a gap with the fewest and lowest obstructions.

We also had to be careful when landing on a seemingly level spot surrounded by mountains which blocked the natural horizon. Once again, a dangerously false impression could result in the landing area being far too steep. Before attempting to carry out a landing, we would turn the helicopter to face up the slope whenever possible, and then slowly lower the collective. The nose wheel should then touch the ground first. But occasionally, the cyclic would still come against the forward 'stop' before the rear wheels had even reached the surface and allow us to complete a safe landing.

On one of our trips, while I flew with one of the students, we left Ken Law with the other one, at the Outward-Bound school. The flying student was then introduced to the exercise of cargo swinging fuel drums and other equipment in and out of LZs in mountainous terrain. While we flew, I impressed on the flight engineer just how important it was to regularly lie on the rear cockpit floor and check the nylon net's underslung load. At the end of this sortie, we returned to land at the Outward-Bound football field, which was at the base of the mountain range near a gully called 'Banana Grove.' As we came over the top of the lower plateau, with 3,000 feet to lose, I could see the other waiting helicopter. We had to carry out a fast rate of descent over the short distance to the football field, nestled among the buildings. As we plummeted in autorotation, with the collective fully down, I listened to the rotor blades' clattering sound, while I noticed someone running towards the other helicopter. Assuming he would call us on the radio, I gave a light-hearted commentary on his progress. However, my heart skipped a beat when I heard Kenny's frantic voice coming over the radio, "Bravo 7, the drum has fallen out of your net, which is nearly touching your tail rotor!" I realised the imminent danger, and gently pulled up on the collective while keeping a level nose attitude, to avoid the empty cargo net snagging the tail-rotor. If that happened, we would lose the tail rotor's counter-rotating effect and start to spin in the opposite direction to the main rotor blades. If that happened, we would have no option but to go into autorotation and hope to get away with a crash landing in the uninviting mountainous terrain. I gingerly continued with a gentle approach while the engineer lay over the edge of the cockpit floor to ensure the net stayed clear of the tail rotor. Taking this precaution, we were able to make a safe landing and once again realised that the Lord above had been watching over us. If Ken had not been there at the right moment to look up or act as quickly as he did, the net could have swung back into the tail rotor when we flared the aircraft on the final stages of the approach and could have culminated in a disastrous outcome.

Our training was mainly over undulating terrain, covered by steep granite slopes and huge boulders strewn over the relatively flatter areas that offered us few places to perform a forced landing in an engine failure event. So, we would impress students to continually keep an eye open for possible landing spots. To emphasise this point, I would periodically push down on the collective to simulate an engine failure and put the helicopter into autorotation at unsuspecting moments. Sometimes I would let the helicopter drop like a brick through a few thousand feet while the student carried out 'S' bends to guide the descent into the most favourable area below. He would also need to keep an eye on the main rotor RPM gauge and pull up the collective to avoid rotor 'over speed' in the thin air of the high-density altitudes in which we operated.

A favourite spot where we enjoyed having our lunch breaks was at Gossamer Falls. This is a most spectacular waterfall deep inside Mozambique. To reach it, we had to fly over dense impenetrable teak forests that stretched over 300 feet above the forest floor. The only access to the area on foot would be by following ancient elephant paths. But, although the falls were visible from a distant point, across a densely treed valley, hikers were never able to reach it on the opposite side where the water fell in a fine spray 800 ft down a sheer rock face, to the river below.

We would land at the top of this waterfall, where water swirled around a smoothly polished granite ledge. There was just enough room to put down one helicopter, where the instructor had to fly the aircraft from the middle seat. While he carefully hovered over the spot and placed the nose wheel in a shallow depression, the engineer hung out the cabin's left side and the student out to the right. They would help the instructor swivel the main wheels onto a suitable position. Once all the wheels were down, the instructor would carefully lower the collective. He would then give it gentle jerks to ensure the helicopter was firmly settled and not about to slide backwards over the falls. It would have been catastrophic if that happened after we shut the engine and the rotor blades had stopped spinning. Once we had gingerly climbed out, it was an impressive and thrilling sight to see the helicopter's tail sticking out over the 800-foot cliff. We would then swim in a pool just upstream, enjoy our sumptuous picnic lunch, and bask in the sun.

I shudder to contemplate the consequences if we had ever struggled to restart the helicopter, or suffered some other serious mishap, especially with no place for a second helicopter to land. We had no authority to penetrate so deep into a foreign country, which could have triggered international repercussions. It would also have been an almost impossible task to walk out of the area.

On other breaks from flight training, there were many beautiful spots and sparkling pools in which we bravely swam in the clear but icy mountain streams. The games department had previously stocked trout into some of the more easily accessible mountain streams, which we had sometimes assisted by bringing fingerlings that we had collected from the Mare Dam hatchery. This was in the beautiful countryside of the Inyanga National Park, just to the west of the Inyangani Mountain near the Rhodes Inyanga Hotel. This magnificent stretch of rolling mountainous expanse had been a few adjoining farms bequeathed to the nation by Cecil John Rhodes. At 8400 feet above sea level, Inyangani Mountain was a two-hour flight to the north of the Chimanimani mountains and the highest spot in Rhodesia.

The two mountainous areas are quite different. The Chimanimani mountains are rugged, towering, majestic vertical rock-faced spines with grassed and sheltered valleys cradled high up between them. These valleys have meandering streams and waterfalls, while the Inyanga mountain area is covered by rolling grassland downs, with introduced forests of pine trees and wattles. It is a beautiful and spectacular 'chocolate box' place and not so challenging or suitable for mountain flying training on helicopters.

I came to love the Chimanimani Mountains with their stunning panoramic views and the opportunity they gave to hone our helicopter flying skills. They familiarised pilots with fantastic diversity and potential excitement when flying helicopters. They also highlighted the vulnerability posed by rapidly changing weather conditions and the challenge of landing in confined spaces with numerous unforgiving obstructions to avoid.

Nearly 30 years later, we lived on a beautiful property at 7,000 feet AMSL in Troutbeck near Mount Nyangani, a continuation of Zimbabwe's Eastern Highlands. Our sweet, crystal-clear drinking water was from a borehole fed by pristine mountain streams filtered into an underground river.

While there, we enjoyed the pleasure of cosy log fires with wood provided by freely accessible and prolific wattle trees that grew like weeds on the mist-covered slopes. I have happy memories of this beautiful mountain range extending along the eastern border of what was once God's Own Land.

19 BOB MACKIE FLIES ME TO MUKUMBURA

L-R Peter Rawlings, Brian Meikle, Brian Phillips, Me, Roy Stewart, Bob Mackie

No 7 Squadron came under tremendous pressure to meet the demands of the escalating conflicts on Operation Hurricane. Helicopter crews were now required at Centenary and Mount Darwin while still supporting the Forward Airfields (FAFs) at Kariba and Wankie. The Air Force had also transferred Mick Greer to HQ, after being our Squadron Commander for two years before the commencement of counter-insurgency operations.

HQ decided to attach some previous serving helicopter pilots to help No 7 Squadron meet the added demands. The Squadron also needed to use the instructors in the operational field, so there was no one left to train the four new pilots posted onto Alouettes. Therefore, these recruits, including S/Ldr Eddie Wilkinson, who had been posted as No 7 Squadron's newly appointed Squadron Commander, were sent to South Africa to undergo helicopter conversion with the SAAF (South African Air Force), which took place at the Ysterplaat base, near Cape Town.

To fill the void, one of the seconded pilots, was S/Ldr Randy Du Rand, who was appointed as the acting Squadron Commander until Eddie Wilkinson's training was complete.

These helicopter pilots quickly formed a special bond between themselves and proudly said they flew as members of the non-existent, No 8 Squadron.

One of my tasks was to carry out refamiliarization and refresher training for the seconded pilots. On one of these sorties, I was busy with Barry Roberts in the Seki general flying area, when we heard Flt Lt Alec Roughead call for take-off in a Canberra aeroplane during the flight. The control tower answered, "Delta Five, cleared for take-off, call turning out left." The subsequent transmissions we heard were:

Alec: "Delta Five, turning out right", ATC: "We instructed you to turn out left!" Alec: "Roger, turning out left!" ATC: "Don't worry. Continue with the right turn!" Alec: "Roger, turning out right!"

The next call to Delta Five from ATC received no response. After further repeated calls, we heard nothing more. The ATC then asked us to go to an area east of the airfield, where they could see a thick column of smoke. When we arrived overhead, we saw the Canberra wreckage and guided a fire jeep to the spot. We them landed close to the wreck, where a witness pointed out where two bodies were lying. But I could not bring myself to go and look.

The navigator, Guy Robertson, and his parents were family friends when I was a teenager, and Alec and I had been on the same cadet training course. He had also been my best man when Tish and I were married, and a few years later, we asked him to be Shane's godfather.

It turned out that the accident had been the result of the main spar on one wing snapping. The possible cause was Alec's banking the Canberra snappily, from one direction to the other, while pulling back on the elevator. Canberra pilots were only allowed to pull 2G when applying aileron in a high bank turn. Sadly, the pilots were honour bound to keep their seat pins in until above one thousand feet, since the navigator's ejector seats could only be activated safely above this height.

During that time, the morale of the Alouette Squadron had taken a massive blow. The wives, unaccustomed to dealing without their husbands about the house, were now taking the strain. Their husbands were spending extended periods away from home, and the wives had to deal with the accounts and other problems on their own.

Randy decided to invite the pilots and wives to a party at his house when he realised just how low the Squadron's morale had become. He also included the wives whose husbands were in the bush. Apart from his Air Force career, Randy also owned a dairy farm near Beatrice. His wife, Gerry, was a gracious hostess who laid on a sumptuous meal for us in true Afrikaans farming fashion.

The party was a real hit when Randy introduced us to downing a shot of Schnapps, after licking salt from between a thumb and forefinger, and then followed by sucking a quarter lemon. Within a short time, we had lost all inhibitions, which gave us the chance to let down our hair and release all built-up tensions. In no time, we ended up playing a game called Dead Ants, where someone would call "Dead Ants", and we would then jump on top of each other in a pile of bodies, with shrieks of laughter. After thoroughly enjoying this bit of fun, the party of *'thirty-somethings'* ended up in the swimming pool.

The party went down so well that we decided to have

a repeat performance at Barry and Shelley Robert's house the next night, which was even more fun than the night before. I consumed far too much alcohol, even though the programme tasked me to deploy with a helicopter to Mukumbura the following day.

Before all this and the increased commitments, stints in the bush were relaxed affairs. So, a few flight engineers would ask if the dual controls could be connected so that they could have some flying practice. Although realising authority would never sanction it, I encouraged their interest as I felt this flying ability could someday prove beneficial. A few engineers became so proficient that I thought their ability surpassed many Squadron pilots. One day, we allowed them to compete against each other in a Squadron hovering competition. The pilots had already enjoyed three of these competitions, in which I had managed to win the last two. After joining the Squadron, what helped me was the steady approach I had witnessed by Air Sub Lt Brian Phillips. I recognised that this was how to fly a helicopter, as pilots would need to hover carefully and smoothly.

There were various tasks that we incorporated into these competitions. One involved cargo swinging a concrete block onto a 200-litre fuel drum, and the other lowering a sandbag attached to the hoist into a pile of tyres from forty feet up, where the hoist operator would winch down the sandbag into the set of four tyres. Monitoring judges would time each exercise from when the helicopter left the ground. Another part of the event included hover-taxiing a cement block through a winding course between pairs of drums. During this exercise, points were deducted each time the cement block touched the ground, struck a drum, or was seen higher than the drum tops. A tricky part of the competition was the spot landing, where the pilot had to place each wheel as near as possible to paint marks on the helicopter pad.

Since I was a qualified helicopter instructor, I acted as

the safety pilot from the middle seat to make it relatively legal. Bob Mackie, Brian Warren, Bernie Collocot, Pete McCabe, and one or two others took part, and most of them achieved better scores than many of the qualified pilots. As an indication and credit to their proficiency, I never needed to touch the controls.

The day after the party at Barry and Shelley's house, I was in no fit state to fly a helicopter, as I was suffering one of the worst hangovers I had ever experienced and barely made it to the Squadron. Thankfully for me, Bob Mackie was my flight engineer. I asked him if he thought he could fly the helicopter to Mukumbura, while I slept on the bench seat behind the front seats. Bob assured me that it would be no problem. With his having to do all the radio calls and navigation, I have no idea how we got there, but he duly delivered the helicopter to Mukumbura that day. An exceedingly sick pilot was fast asleep on the uncomfortable back seat when it arrived.

Fortunately, there was not much going on that day that needed my attention. So, I sat quietly in a shady spot to keep out of the blazing sun. At about midday, Giles Porter arrived with a ferry aircraft for me to return to New Sarum. The flight commander told him to take over from me, as I was to leave the next day for Cape Town. Once there, I made my way to Ysterplaat and reported to Maj Gert Van Zyl. He was the Squadron Commander, and our new pilots were busy with helicopter training. The SAAF asked him to give me a short conversion onto the Alouette II, the Alouette III forerunner, as they had donated six of these helicopters to the Rhodesian Air Force. With the ever-increasing operational demand, we would later use them to supplement our inadequate number of Alouette IIIs, once we received these new aircraft. We later used them for initial introductory helicopter training and crew changeovers in the field.

When I arrived at Ysterplaat, I was amazed at the lack of progress made by our four trainee pilots, who had been there for nearly three months. I was surprised by this until I discovered the complete lack of urgency by the SAAF training Squadron to churn out helicopter pilots. The SAAF also had very limiting weather conditions in which students were allowed to fly. For example, once the wind speed reached ten knots, they cancelled all training flights for the day. Since Cape Town is a renowned windy city, this left limited periods when any training was possible.

The holiday atmosphere was immediately evident when the Major responsible for my Alouette II conversion considered that one flight a day was as much as he was willing to do. I familiarised myself with the pilot's handbook and passed the technical exam. Since the powers deemed that I should receive ten hours of instruction to achieve adequate competency, I soon became frustrated with the slow pace of progress. Although the baby Alouette had no powered controls, I was entirely at home in it and felt I needed no further assistance after one sortie. So, after managing only six hours in the space of three weeks, I became so depressed that I suggested we fly two more hours straight off and call it a day.

Unfortunately, Dave Rowe and I had been out for a seafood restaurant meal the night before. Forgetting I had previously suffered seafood poisoning on a trip to Lourenco Marques, I indulged in a delicious crayfish-Thermidor.

The following day, I suffered the effects of seafood poisoning. During the pre-flight inspection, when I looked up at one of the rotor-blade tips, waves of nausea flooded through me, and I was almost sick on the spot. Somehow, I battled through the exercise. Realising that I had already persuaded Gert to fly a two-hour sortie to complete my training, after the first hour, I mentioned that I was not feeling well and suggested we call it quits.

As this was his last chance to fly the Alouette II, which the SAAF had pulled out of 'mothballs' after many years, he asked if he could have a short time of nostalgia at the controls to reminisce and say goodbye. I had no choice in the matter but was ill-prepared for what was to follow. He spent the next fifteen minutes giving a farewell display for the benefit of the onlooking engineers, as he buzzed between the hangars and pulled up into steep torque turns. I just gritted my teeth, praying that I would not embarrass myself by hurling over the shaking instrument panel.

The next day I returned to Rhodesia, and once the six Alouette IIs were delivered and refurbished on No 7 Squadron, I set about giving each Squadron pilot a quick conversion.

After three and a half months, Eddie Wilkinson and the other three pilots returned from South Africa. Despite the time it took to do their basic training, we still had to give them a quick operational conversion, including mountain flying, trooping drills, navigation exercises, and confined area work before releasing them into operations.

Soon after taking over as the Squadron Commander, Eddie introduced us to a formal ladies dining-in-night in the New Sarum Officers' Mess 'Ladies Room'. He asked us to wear our traditional mess kit and asked the ladies to dress in long evening gowns. The evening started tastefully enough. But Eddie was unaware of the close relationships that had developed between pilots and wives. There had been many wild parties following on from the first at Randy Durant's house. Now, after a copious amount of the excellent Cape wine that the four trainee pilots had brought from South Africa, the party quickly deteriorated. The wine had been bought at bargain prices from the officer's mess pub at Ysterplaat, and they had come back with Nederburg Cabernet Sauvignon, which the SAAF pilots referred to as *"Os Bloed"* (Ox Blood) and other high-quality rose and white wines.

Wilky, as his friends in the Air Force generally called him, was unaware of our recent wild parties. The function quickly turned into an event never previously experienced at a ladies-dining-in-night. The wives quickly lost all inhibition and started dancing on the tables, between courses. One senior wife, who shall remain nameless, had so much wine that she flopped out in the adjoining lady's toilet. Mary MacGregor was sitting opposite me, and I am ashamed to confess that I crawled under the table to bite her shin. She got such a fright that she let out a scream and flung herself backwards, which broke the chair.

The new Squadron Commander was so horrified and taken aback by this uncouth display that he vowed never again to have a cultured evening with this bunch of "plebs".

20 HANG GLIDER'S NARROW ESCAPE OVER DURBAN BLUFF

Impala over Durban

In the 1970s, the South Africa Atlas Aircraft Corporation, which fell under the arms procurement Armscor company, assembled the Italian Aermacchi MB-326 jet trainer aircraft under licence. The Rhodesian government, which had been suffering from the imposed punitive UN sanctions instigated by the British Government, desperately needed to replace the ageing Vampire T11 jet trainer, which had served the Rhodesian Air Force admirably since the 1950s. So, they ordered and purchased the first ten aircraft, now named Impala, from South Africa.

Unfortunately, before South Africa could deliver these ten aircraft, the U.S. Secretary of State, Henry Kissinger, met with the South African Prime Minister,

John Vorster, the Zambian president, Kenneth Kaunda, and the Rhodesian Prime Minister, Ian Smith. This meeting's outcome was to induce John Vorster to withdraw further support and assistance to the Rhodesian war effort against insurgents.

Although this happened when the Rhodesian armed forces had just about brought the situation under control, the loss of aid from the friendly neighbour to the south, posed a severe setback. To complicate matters, South African had to cancel delivery of the purchased Impala aircraft. So, as a private goodwill act, they officially recruited the Rhodesian instructors and trainee officer cadets into the South African Air Force, which was carried out in such a clandestine way that most of our Air Force personnel knew nothing about it.

The first few courses operated out of the South African Air Force training base at Langebaanweg, north of Cape Town, with Squadron Leader Hugh Slatter as the Squadron Commander. This location proved quite a challenge since most Rhodesian Air Force pilots were English-speaking, and not all of them had taken Afrikaans as a secondary language at school.

The SAAF had the policy to alternate the languages each month for all correspondence and official business. In theory, they applied this rule, but as most pilots in the SAAF were Afrikaans speaking, they did not always strictly adhere to it. Trying to conform to this policy presented some serious flight safety problems for our pilots, especially when the Mirage fighter Squadron arrived to conduct air-to-air target practice from the same base.

The solution was to move us to Natal, a predominantly English-speaking area. They decided to shift, what was known as "Charlie Flight," to the sleepy

Air Force base at Louis Botha Airport, which lay a short distance to the south of Durban.

Meanwhile, in Rhodesia, I had spent six years on No 7 Squadron, operating the Alouette Helicopter, before being promoted to the rank of Sqn Ldr and posted to Air Force HQ. A few months later, I was overjoyed when I received the plum posting to Durban, to take over from Hugh, who was promoted to the rank of Wing Commander, and became OC Flying at New Sarum, back in Rhodesia.

Durban is one of South Africa's major seaside holiday resorts. High temperatures and the warm Indian Ocean current create a languorous atmosphere, to the extent that the British colonial style, military headquarters, not far from Durban's most popular beaches and officially named Natal Command, is referred to as *"Holiday Command"*, due to the laid-back attitude towards any productive work.

Most holidaymakers, who poured down to the coast in their droves over Christmas and Easter, were Afrikaans-speaking people from Transvaal, and the historical enemies of the people of Natal. They had been fed stories about the atrocities inflicted upon them during the Boer War, seventy years before. The locals referred to these visitors in a derogatory fashion, as "Rocks". So, when they flooded the beaches over Easter and other holidays, the influx was called an "Avalanche".

Given the rank of Major in the South African Air Force, I travelled to South Africa on the Affretair DC7 with Number 30 PTC cadets. They had never met me and were unaware that I was their new Squadron Commander.

We were first sent to Langebaanweg for a week to undergo simulator training. On the second day, I was embarrassed to witness the casual way our squad of officer cadets marched from the officers' mess to the security area. When I saw their heads swivelling around to look at every Mirage flying overhead, I realised that I had to do something about it before their Wings Parade in about seven months. Once they arrived in Durban, I made a rule that from then on, wherever they went about the station, they had to march like clockwork soldiers with their arms swinging shoulder high. The apathy that pervaded the Air Force base made the cadets' energetic marching seem like a breath of fresh air. But I am sure the students did not see it in the same light.

The cadets stayed at the Navy Mess on the Bluff, which was a high coastal ridge east of Durban harbour. The instructors and married technicians lived in holiday flats near the beach at Amanzimtoti for the next seven months, a favourite holiday town south of the airfield. The unmarried technicians, who also started at the same Naval Station, were later transferred to another location to the west of the coastal freeway.

One evening, the instructors arranged to meet the Squadron technicians for a few drinks at the Elangeni Hotel, which overlooked the Marine Parade in the city. While at the round bar, someone introduced us to a drink called *'Poofie Juice'*. It was quite an innocuous tasting drink with a mixture of gin, orange juice, and some liqueur. I had offered to take the technicians back to their accommodation, but as I stepped onto the pavement, the moment I felt the moist sea breeze on my face, my lips went numb, and I knew I had consumed too much.

At that time, I had a sporty VW Variant, and after dropping off my passengers, I had some difficulty finding the South Coast freeway. It must have been quite late, as there was little traffic, so I just put my foot flat on the accelerator to get to bed as soon as possible. I have no idea what speed I reached, but the following day on the crew bus, I faced a chilly reception from Barrie York and Ian McKenzie, who also lived at Amanzimtoti. The night before, they had run out of petrol about ten kilometres before the turnoff to Amanzimtoti, but they were unconcerned as they realised, I would be coming along soon. After quite a while, they heard the roar of my approaching car and stepped into the road to wave me down. However, they had to leap out of the way when I flew right past them at breakneck speed. They must have cursed me as they walked the long ten kilometres back to their accommodation.

The Air Force base housed our ten aircraft, and ten technicians were in the same hangar, and office accommodation, as the SAAF's No 5 Squadron. It was a regular squadron, with over thirty Impalas and nearly thirty-five technicians. With the same laid-back Natal disease, they were lucky if they managed to fly four aircraft once on any given day. On the other hand, with W/O Barrie York the first year, and Rob Rix the second, our technicians managed to have seven aircraft flying four sorties every day. They went on like this for nearly seven months, while still doing complete periodic services and repairs. The local aircraft mechanics inevitably viewed them with undisguised resentment, but in time, we managed to use their junior staff for time-consuming, but straightforward duties. These included marshalling, attending to the aircraft chocks, attaching, removing the cockpit ladders, refuelling, assisting pilots strapping in and getting out of the cockpit, and helping with the ejector seat pins.

The instructors and cadets shared No 19 Squadron's offices, which operated the French-built, twin jet-engine Puma helicopters. They were a wild and boisterous bunch of fun-loving pilots, especially Fritz Picksma, who had played rugby for both Southwest Africa and Natal, and Karl Volker, who had been a junior Springbok gymnastic champion. I had previously met many of these pilots when they had come to Rhodesia to assist us in the anti-terrorist operations.

At that time, Fritz played the Natal team's eighth-man position in the Currie Cup provincial competition. Incredibly fit and an extrovert, he was already a legend in the SAAF and loved playing practical jokes. On a previous squadron, I remember one story about him when he put his hand into the G-suit slit to the side of the hip and stuck his index finger out of the bottom of the full-length front zip. When he walked into the ops room, an innocent young SAAF Ops Assistant girl saw him and cried out in her sweet rural accent, "Ag no Captain, put it away, man!"

Karl Volker was also an exhibitionist. One of his party tricks was to shed his trousers and underpants and do 'flick flacks', which was a sight to behold! The other trick he often performed was to attract a lot of attention when at a party in an upstairs flat. Once all eyes were on him, he would run full speed to dive over the balcony's parapet wall, where his flight engineer would catch him just before he flew over the parapet.

At one party, just as he started his run, someone distracted the now 'sloshed' engineer. So, Karl flew over the balcony without his pants, and landed with his head in a garden pot. Surprised by this unexpected appearance, a passing pedestrian promptly whipped him off to casualty. In the meantime, his engineer wandered around in a *'dwaal'* (daze), slurring to each person he met, "I've lost my pilot", and took nearly an hour to make his way downstairs. Thankfully the accident did not have any long-lasting effect on Karl.

When we were in the No 19 Squadron offices, we often heard the noise created by Ray Houghton, Karl, and Fritz, when they made the most outlandish foghorn sounds with their throats, as they played cricket in the office corridors.

Another Puma pilot on No 19 Squadron was also a character, Denzyl White. He was a quietly spoken, handsome man with dark hair, who was always immaculately dressed and the epitome of sartorial elegance. One could not help viewing him with a certain amount of awe and admiration, as somehow, he always arranged to be the standby helicopter rescue pilot whenever there was a formal cocktail party. He would then organise for the Operation Centre to give him a test call on his 'Beeper' while he was at the function. At the appropriate time, he would have strategically positioned himself near a group of young ladies. The radio call would always stimulate substantial attention and interest from them. He was, however, a very professional pilot and had undertaken some well-publicised sea rescues. Taking himself so seriously, stories flourished about the things he did, such as his going to the bank dressed in his Air Force uniform, and once there, taking his chequebook out of a smart leather briefcase, and handing a cheque to the teller, while wearing a pair of unspoiled pigskin flying gloves.

Then came the day when the *Esmeralda*, a Chilean Naval training ship, sailed into Durban Harbour. A flotilla of yachts from all around South Africa celebrated this spectacular occurrence by meeting and escorting this famous sailing ship, to the harbour mouth. The No 19 Squadron pilots decided to get front-row seats by flying out to the harbour to witness this iconic vessel's historic arrival. Our instructors and cadets were invited to join them for a ride in the Puma to see this event, with

Denzyl White and Fritz Picksma at the controls.

Having spent six years on helicopters and fully aware of the downwash from helicopter blades, I was horrified when we reached the area. The flying pilot, Fritz, proceeded to do low-level 'beat ups,' past the colourful armada of yachts, and as we passed alongside each sailing vessel, I looked backwards out of the open sliding door and was aghast when I saw the downwash blowing over each boat with such force that they almost capsized.

If this was not bad enough, on the helicopter's return flight to base, which was flown south, parallel to the coastline, a few miles out to sea, a hang-glider was spotted. It was just off the Bluff and heading in the same direction. With delight, and to the utter dismay of its pilot, our intrepid crew swung the Puma around and moved into a close formation position on the hang glider's port side. The terrified pilot looked like a fluffy dog, with its head sticking out of the window of a speeding vehicle. While gripping the control bar in one hand, he tried to 'shoo away' this noisy and menacing bulk of a camouflaged helicopter, with the other.

While the helicopter crew roared with laughter, I became concerned about the imminent risk to this flimsy craft. As soon as the helicopter pulled away in a climbing turn to the port, I looked back anxiously, and was just in time to see the downwash from the helicopter's blades flip the hang-glider right over and send it into an uncontrollable spin. In blind fury, I shot into the cockpit door, screaming, "You idiots, you have killed that guy!" My outburst put an immediate dampener on their fun. Their faces instantaneously transformed from hilarious amusement to ashen as they whipped the aircraft around to see why I was shouting. We then caught sight of the distorted shape of the hang-glider's canopy, covering thick bushes at the top of the bluff and the bedraggled pilot standing beside it, furiously shaking his fist at us. The comical sight and relief were too much to hold the crew back any longer, and once more, everyone went into uncontrollable fits of laughter.

As we disembarked and thanked them for the excursion, I could not help admonishing the delinquent crew for their irresponsible display. However, the pilots shrugged it off with, "They are not supposed to fly along the Bluff anyway!" To this, I responded, "So if you see an old lady crossing the road illegally, do you just run her over?" But somehow, this failed to make an impression on them, and they remained unrepentant!

Sadly, when I returned for the next AFS course the following year, I was shocked to learn that Fritz, who had always been so full of life, had joined Court Helicopters and tragically, had been killed when he flew into cables on Cato Ridge, east of Durban.

21 CROP SPRAYING IN RHODESIA

Bell 47 Helicopter with spraying booms

I had just completed a further twelve years in the Rhodesian Air Force.

For the last two years I had been on a top-secret assignment, based in Durban and I felt after this posting, that I could not return to a mundane life. So, on a whim, I decided that crop spraying would be more exciting.

Added to that, the Managing Director of Agricair was Mike Saunders, who I had so admired when he was the Squadron Commander of No 1 Squadron. During those heady days, I was flying as a 'Jet Jockey' on the newly acquired Hawker Hunters.

As a twenty-one-year-old, I had become easily

influenced by and admired the dashing style of the senior pilots on that Squadron. They were a wild bunch who took pride in their flying but let their hair down at parties. As the Squadron Commander, Mike was a bright, suave, and handsome man, with a style of his own, who appealed to my impressionable mind. At that time, stories abounded about his previous escapades. The one I particularly enjoyed was when he had been the *aide-de-camp* to Lord Dalhousie, the Governor-General of the Federation of Rhodesia and Nyasaland. Mike fitted the bill at face value, but I believe this posting ended abruptly when the Governor's wife caught him hurling her pet cat down the stairs.

Now, years later, at thirty-seven, I still held Mike in high esteem. When he suggested I would be an asset to Agricair, it seemed a good time to make a career move, especially as Mike thought my years of instruction would greatly benefit his organisation. It also fascinated me that they operated several crop-spraying aircraft as far afield as Sudan.

Agricair based its operation at Charles Prince Airport, north of Salisbury (Harare), and Mike managed it in much the same way he had run the Squadron many years before.

Before the end of my training as a crop spraying pilot on helicopters, I was involved in a crash that nearly killed me. That day, I went on a training session with George Unger, a Norwegian, who was the chief pilot. Although he was a qualified instructor on the Bell 47, he had never actually done crop-spraying from helicopters. Each year, he only flew the minimum hours required to keep his licence current.

On this sortie, we took off from Charles Prince Airport and flew to an open field off to the east. Here, he showed me how to do crop spraying turns at the end of each run, where the helicopter was expeditiously turned around and lined up for the next run. It was an effective manoeuvre when the aircraft was light. But when the plane was fully loaded, it took some practice to perfect. The most effective technique was somewhere between a torque turn and a regular turn, depending on the load.

A torque turn is like a stall turn, one of the aerobatic manoeuvres in fixed-wing aircraft. During this, the pilot pulls the plane up into a vertical climb. Then, as the speed decreases, the pilot fully applies the rudder to put the craft into a flat turn around its vertical axis, and the manoeuvre takes some skill to ensure that the aircraft is kept under control. When done in a helicopter, the collective is also brought into play to assist the turn. When the turn was in the same direction as the torque, the pilot would keep the collective up, but it would be lowered when the turn was in the same direction as the rotor blades.

After an hour of buzzing up and down the chosen field, I felt comfortable with the manoeuvre. So, we headed back to the airfield at about 300 feet above the ground. En route, the helicopter started to slew around in the opposite direction to the rotor blades. I first tried to counter the rapidly increasing spin rate by applying the opposite yaw pedal, but quickly realised it was due to some failure of the tail rotor system, which opposes the engine's torque effect.

As George was the training captain on this flight, I released the flight controls to him. But he surprised me when he failed to put the helicopter directly into autorotation to remove the torque. Instead, he tried to maintain the aircraft in level flight while revving the throttle. His action only aggravated the situation, and in no time at all, we were in a flat spin, and the aircraft started to wobble furiously from side to side.

On many previous occasions, I had trained helicopter students for such an emergency, where I would first bring the helicopter into a hover at 500 feet above the runway, and then cut the engine. When this happened, to maintain the rotor RPM, the student had to put the aircraft into autorotation by immediately 'dumping' the collective. Simultaneously, he would have to lower the nose to allow the forward airspeed to build up enough to flare into level flight, just before touchdown. At this stage, with the undercarriage just above the ground, the pilot would progressively pull up on the collective. This would use up the rotor's kinetic energy to cushion the aircraft into a soft, roll-on landing. Anything below 500 feet was known as the helicopter being in the 'dead man's curve', as designers could not guarantee a safe landing beneath this height.

Now, when we lost the tail rotor function, we were precisely in this danger zone, and had lost all forward speed by failing to take the immediate corrective action. Straight away, I realised the difficult predicament we were in, and was sure it was moments before being killed.

I can still recall the thoughts that ran through my mind. Time changed into slow motion when I knew this was seconds before being killed in the impending crash. By that time, I had lost many friends in aircraft accidents, and I knew just how some of them must have felt. I also thought of my wife and felt sad when I thought she would never know the exact circumstances surrounding my death. Furthermore, I was angry with George for not carrying out the correct emergency procedure. I later learnt that George had been thinking of the laundry he had not collected from the launderette!

After having spent many hours flying as an instructor, I responded by a conditioned reflex action, to snatch the controls from George and dump the collective. But unfortunately, there was insufficient height to build up any forward speed, the rotor RPM had also wound right down, and we did not have enough height AGL, to give it a chance to recover. The ground rushed up towards us like a freight train, and at the last moment, I pulled full collective to cushion the landing. However, this did little to slow the downward plummet towards earth, and we impacted the ground in a whirl of dust and debris.

I was surprised to find that I was still alive and shut the engine down before dragging myself painfully out of the wreckage. The helicopter, lying there in a mangled heap was a complete write-off. The force of the impact had caused my head to lurch forward, and the helmet hit the cyclic plastic grip so hard that it snapped right off in my hand. As I climbed out of the sad-looking wreckage, I felt as though someone had kicked me in my solar plexus, and my back was also excruciatingly painful.

Not thirty metres from where the aircraft had come down with a bone-jarring thud and cloud of dust, an African woman was busy hoeing the ground. Amazingly, she just continued to work without an upward glance. As I staggered away from the accident, two young African boys appeared out of the long grass at the edge of the field. As things had happened so fast, we had no chance to transmit a distress call during the emergency. So, I asked these youngsters to take a scribbled note to one of the houses in a nearby suburb. I had just started writing the message to ask the recipient to phone Agricair, by using the telephone number I had just included, when a woman and her son appeared. They had seen us coming down and located us amongst the mealie field and kindly offered to take us back to Charles Prince Airport.

I kept the jotted down note, but although I thought I was thinking quite clearly and rationally at that time, when I found it in my pocket, I was amazed to see that the telephone number I had written, bore no resemblance whatsoever to that of Agricair's.

When we walked into Mike Saunders' office, George Unger threw himself onto the carpet with a dramatic flourish. As his bone dome (crash helmet) went rolling across the floor, he stated, "I'm dead". As the onlookers stood in stunned disbelief and confusion, I had to activate someone to ring my wife, and take us to hospital.

Later, as I lay in the waiting area in casualty, after having had an x-ray of my spine taken, I was alarmed when the attending doctor called some of his colleagues to come and have a look at my X-ray plate. They paid no attention to me as I lay there in agony, while their attention was transfixed on the x-ray, until I croaked, "Is something wrong?" The attending doctor apologised as he turned to reassure me. He explained that what had surprised them to see, was that there was an extra vertebra in my neck, which none of them had ever seen before.

During the flight, I had worn a shoulder harness. On the other hand, a lap strap was the only thing holding George in the cockpit. I am sure this accounted for the differences in our injuries. He suffered a cracked vertebra, while I had a compression fracture to my spine. For months after this, I experienced severe pain behind my shoulder blades which caused considerable discomfort, especially when I applied shampoo in the shower.

In the meantime, my brother-in-law Dave Thorne had driven to our home to alert Tish about the accident. She then came straight to the General Hospital to collect me. As she went along Moffatt Street, and was turning left into North Avenue, I put my head out of the window and shouted to all and sundry, "I'm alive! I'm alive!"

The accident shook my confidence. Particularly when the next flight on the Bell 47 was only ten days later, and was my first real, crop spraying job. I also had to take off and fly with a full chemical load in the hoppers. It was a shaky performance, as I was too nervous about pulling the helicopter around at the end of each run in a sharp turn. The blades also made a loud clattering sound as they struggled against the extra weight. My right leg shook uncontrollably the whole time, and I hyperventilated so severely that I was in danger of passing out.

Over the next few weeks, I regained some composure, but came to realise how much the Air Force had cast me in its mould. It was difficult for me to relate to the 'gung-ho' cowboy mentality of the other crop-spraying pilots. They were a friendly bunch, but I soon discovered they were a wild lot, who came from all walks of life and were quite different from Air Force pilots. This was a rude awakening, as I came from such a structured, military background. I now had to relate to a completely different calibre of undisciplined pilots. To my mind, although they were good at their job, they lacked a professional approach to their flying.

Although a few technicians had come through the Air Force, only one other pilot, apart from Mike Saunders, had served in the Air Force.

Of course, there had to be a pub where the pilots and staff could meet after work or check in for a few drinks whenever they passed through the base. This was done between stints in the farming areas, or when they returned from outside the country.

Sensing my attitude towards their undisciplined approach to flying, some went as far as to tell me that my Air Force training and background could never adequately prepare me to become a real *'bush pilot'*. Especially as I could not drink all night and then fly from first light without any ill effects. They had an opportunity to drive this point home when the company sent me to Glendale for my first crop-spraying job. I arrived the day before to spend a night in the crew house. With hardly enough time to settle in, the technician whisked me off to a nearby farm dam. Once there, I watched in absolute horror as two aircraft crisscrossed each other as they skied across the surface. Stuart flew the streamlined Piper Pawnee, while Mike flew the chunky Thrush Commander, with its ring of throaty cylinders on the bulky radial engine.

I must confess, I could not help admiring their flying skill as the wheels aquaplaned across the surface of the dam, sending plumes of water arching behind each aircraft. Although it was an astonishing display of bravado, I was still horrified by their haphazard synchronisation. Especially as they continually crossed each other's paths from random directions. I could just picture an accident waiting to happen. Although I had taken numerous chances in my career, I still felt it was the most irresponsible demonstration of bad airmanship I had ever seen.

Adding to my discomfort, once we returned to the house, I had an upset stomach and was in no mood to spend any time drinking at the country club.

From experience, I knew that if I spent the night propping up the bar, I would not be able to fly the next day. In any case, this contravened all regulations, which require twelve hours between 'bottle and throttle.'

Therefore, I excused myself early, facing much scorn and ridicule, and went off to get a decent night's sleep. But I was soon disturbed by a squadron of mosquitoes whining around my head. Later, I was tipped out of my bed when the pilots and technicians returned from the pub. I shared a room with Mike that night, but once the frivolities had ceased, and in a state of intoxication, Mike finally collapsed on his bed, and then his loud snoring ruined the remaining few hours.

I am sure they felt justified in their attitude and successfully proved that Air Force pilots were 'wimps.' It also showed that my rank of Sqn Ldr, meant nothing in this environment. On the other hand, it confirmed that I had made another grave mistake, assuming I would enjoy this career change.

I had grown up in a disciplined organisation in the Air Force, where I felt like a valuable member. It was also small enough to feel like part of a close family where I felt accepted, with all my strengths and weaknesses. I also felt recognised as someone who had contributed as a qualified instructor on both fixed-wing and helicopters. The Air Force also qualified me as an air weapons instructor and instrument rating examiner.

Now, it was a humbling experience in this new environment, and my previous experience counted for nothing. I would have to stand on my merit, devoid of the spurious sense of worth as an Air Force officer.

However, despite my misgivings and regret, I became proficient enough in this job. I achieved the same number of hectares per flying hour as Dave Thorn, an accomplished helicopter spraying pilot, who had previously flown helicopters in the Fleet Air Arm and was now employed by Agricair. I first met him when I was a flight commander on No 7 Squadron, and he had joined us as a direct entry pilot. However, I always found his approach to flying rather unconventional, but was grateful when he later passed on many crop spraying tips that helped me. Nevertheless, this job was not as exciting as I had expected; instead, it was hot, grinding, and monotonous work.

Although I had accumulated over 2,500 hours on the Alouette helicopter and had done over 500 hours as an instructor, the Bell 47 was quite different. For a start, it was piston driven by a 140 horsepower Lycoming engine, which did not have the luxury of a fuel governor, which automatically keeps the rotor RPM in a safe flying range. Instead, the pilot had to manually control the rotor RPM, using a motorbike-style throttle grip on the collective.

This throttle grip gave the pilot one more thing to worry about, as he had to continually monitor and control the RPM within a tight bracket for safe flight. As the helicopter staggered off in *'ground-effect,'* the throttle had to be increased progressively to avoid rotor 'decay' and performance loss. At the lift-off point, the pilot had to increase the engine revs to the maximum allowed, before yanking up the collective to claw its way into the air.

My allocated area of operation was in Shamva, a farming area to the country's northeast. It was an area where Don Northcroft, a pilot I previously mentioned who had left the Air Force to join his father, and work on his farm there. He was an extremely talented sportsman, who had left Prince Edward School as the Head Boy. Their farm bordered a tribal trust land, which had been infiltrated by terrorists. They had learnt that Don was an Air Force pilot, who could be called up for counter-terrorist operations. So, they took the opportunity to ambush and kill him, while he was busy inspecting their lands on his off-road bike.

Although terrorists surrounded the farming area and watched the activity, they never interfered with crop-spraying aircraft. Every weekend I would drive home in a company car. It took just over an hour to get to Salisbury, and I knew there was always the risk of facing an ambush, as a few had already taken place on the same road. So, to feel more secure, I always drove and flew the helicopter with a Mosburg pump-action shotgun beside me.

During that time, I shared the farming area with an ageing Polish pilot called Tad. He was only fifty-four years old but seemed ancient as I was thirty-six at that time. Sadly, I offended him one day after he asked if I knew how to play chess. Although I had studied the game extensively and played for many hours while sitting around in the bush on Air Force operations, I pretended that I only had a vague knowledge of the game. We agreed to play when I returned to the house after spending the weekend at home with my family. Farmers in the area had not booked me to spray the following Monday, so when I returned later that afternoon, Tad was waiting for me, with a chessboard set up in front of him. It was stinking hot after my drive and so I said I would need a shower and a cup of tea before joining him. He had left the real chessboard at home but taken time to draw a magnificent replica on a piece of cardboard.

When we finally settled down to play, I asked him to remind me how the various chess pieces moved. I had learnt certain moves where I could gain some advantage by swapping pieces with similar values. So, when I made a hesitant aggressive move as though I had no idea what I was doing, he looked a bit bemused but could not see where I was heading. I then tempted away one of the pieces protecting his King by placing one of my chess pieces in a position where he could take it for no loss. I then moved into a place where he could not block my next move to checkmate, and I never looked up as I tentatively took up this position. He did not realise what was coming and moved another one of his pieces. I took a protracted period while pretending that I did not know what to do next but kept looking down. I picked up the moment he realised the situation by moving his head up and down sharply, looking at me and then back at the board. I pretended I did not know where to move until I finally asked uncertainly, "Is this checkmate?" This led him to tip over the small table and storm out of the house for a few hours. I felt guilty afterwards, as he took it so badly that he never spoke to me again.

The other person who shared the house with us was John Britton, who had been a flight engineer on Alouettes when I was on the Squadron. I had always known him as a powerful man with a fantastic sense of humour, who had kept us entertained for hours and I recalled the numerous occasions he had put on an act. I was so pleased that he was in the house to counter the atmosphere that Tad felt towards me, and we would often take the opportunity to play squash on the smooth, dusty cement floor at the sports club court.

Within the first week, I fell into a deep depression

when I realised that I had made another terrible and irreversible mistake by leaving behind my happy Air Force life. I even approached Mick Greer, my previous Squadron Commander on No 7 Squadron, to ask if I could rejoin the Air Force. My spirit instantly soared when he informed me that A/Cdre Norman Walsh was happy to take me back. It felt like a heavy suffocating blanket had lifted off my shoulders. However, the next day, my hopes were dashed when I learnt that the added proviso would be to take a drop in rank. It meant that I would have to drop back to the Air Force rank of Flight Lieutenant, when I had been on the verge of being promoted to Wing Commander, a few months earlier. Sadly, my pride did not allow me to the humiliation and embarrassment of returning under those conditions. So, I had to let go of any dream to take up where I had left off.

During that season, I got by with about ten per cent concentration in the sweltering hot cockpit, while my mind grappled with what alternative prospects the future had to offer. I finally concluded that I should join some ex-Air Force friends flying for Court Helicopters in Cape Town. As I buzzed up and down the cotton fields and ducked under electric wires, this decision helped me get by. Ignoring the baking heat and endless sweat, I just held onto a mental picture of the cold, cobalt-blue sea, off Cape Town.

One of my clients was Aubrey Logan, who had married Edone-Ann, an elder sister of John Petheram, who had been one of my best friends at Prince Edward School. On one occasion, when I was spraying their cotton fields, one end of the land was up against a large granite outcrop. When I had to do a torque turn over one of these round rocks, I did not realise that I would be turning downwind. As I was dropping sharply back onto the field, I felt the aircraft shake in a rapidly increasing rate of descent. I immediately guessed that I was in what is called 'A Vortex Ring', which is a bit like undergoing a stall in a fixed-wing aircraft. In the Air Force, we had always made sure we avoided this situation by going into level flight before descending for a landing. This method avoided the helicopter from sinking at the same rate as the downwash from the main rotor blades. As I could see the ground rushing up at me, the automatic response would usually be to pull back and lift the collective to climb away. But something prompted me to do the opposite thing, and I lowered the collective lever and pushed forward on the cyclic. After a slight shake, I felt the helicopter blades bite into the air again and managed to level off just before striking the ground. I was so grateful for the years I had spent lecturing and instructing on helicopters, which stored this knowledge somewhere in the back of my mind.

When the season finished, I submitted my resignation from Agricair. By then, Mike Saunders realised just how out of place I felt in this flying environment and that I just could not fit in with this wild and hard drinking, but skilled bunch of aviators, who performed so admirably in the high-risk but lucrative job.

Unbeknown to me then, I would later be asked to establish a crop spraying division for Court Helicopters in the Western Cape!

22 OPERATION NEAR MONTE CASINO – MOZAMBIQUE

Deploying troops from the hover

In August 1978, after completing another twelve years of service, I retired from the Air Force and joined Agricair, a crop spraying outfit based at Charles Prince Airport. I had spent six years flying Alouette helicopters and then two years based in South Africa as the Training Squadron Commander.

By the end of 1972, I had been on No 7 Squadron flying Alouettes for two years before the 'bush war' flared up in earnest. Because of this, we had been in an ideal position to become intimately involved in the development of new strategies to combat the terrorist insurgency. At that time, there was a nucleus of pilots and flight engineers on the Squadron, who had become a close-knit team. We were young, invulnerable, fearless, professional, and dedicated to the task.

Now, seven years later, every available civilian was being called up for extended periods to do a 'stint' in the bush. I was busy with crop spraying cotton in the Shamva area when I received a call-up.

I had been away from operational flying for nearly three years, and now I joined a completely different set of helicopter pilots, many of whom had even been students when I was the Squadron Commander in Durban. Now, in a complete turnaround from previous Air Force policy, they had been posted to No 7 Squadron shortly after receiving their wings, and only recently qualified as operational helicopter pilots.

What was immediately apparent, was the lack of cohesive teamwork. During my years on helicopter operations, the close cooperation that we had previously enjoyed was no longer there. Procedures had since changed, and I now felt completely out of touch. The other thing missing amongst this new set of pilots and engineers was the trust and understanding we had known such a short time ago. On the other hand, I am prone to think they regarded me as a fossil from the past.

There was no doubt that the vibe and intensity of the bush war had changed. The security forces were no longer on top of the situation, and incidents spread like an out-of-control cancerous growth as it continued to engulf the country. Political pressure and economic sanctions had also taken their toll. With the continuing conflict, it now cost the government one million dollars per day. Civilians were also spending up to two months on call-up duties, as the war had stepped up a gear, and I now felt completely out of tune with the current situation.

Being the 'new boy on the block,' I was assigned the 'tail-end Charlie' position whenever the fire force responded to an ever-increasing number of callouts. The troops, now employed on fire force duties, were no longer the crack RLI, but most often, the civilians on call-up duty. The situation had also become so chaotic, that I was baffled most of the time. To add to this, nobody was told what the problem was, which only exacerbated the confusion.

On the first contact after arriving, I had a flight engineer, who I had never previously met, and who wore contact lenses. After reaching the scene, I spotted a terrorist, lying in camouflage on his back under some trees and pointing an AK47 assault weapon at us. In the meantime, I was still waiting to deploy my 'stick' of *troopies*, and went into an orbit at 100 feet, while I flew at 45 knots around the terrorist's position. I tried unsuccessfully to draw my engineer's attention to the terrorist's location. But becoming exasperated, I instructed him to open fire in the general area with his MAG and hoped this might help me re-direct his next shots. Then, as the flight engineer opened fire, the terrorist, who had been keeping an eye on us, and kept pointing his weapon towards us, answered with a prolonged burst of automatic fire. At that point, we were less than 50 metres apart, and it sounded like a swarm of bees whistling through the open cockpit. Rounds pierced the windscreen, and grazed my engineer's right shoulder, as it penetrated the MAG chest plate. A piece of shrapnel also struck just below my right eye, but miraculously, none of the rounds fired from that close range, hit anyone else in the cockpit. With the flight engineer out of action, I had no option but to pull away. So, after dropping off my stick, I returned to base, feeling utterly disheartened by the lack of cohesion I had just experienced.

After a few months, in October 1979, I was on call-up again. Once more, I was at the back of six helicopters positioning from Grand Reef Airport near Umtali to Lake Alexander in the Inyanga district.

En route, I was shocked at how low the helicopters ahead of me were flying. Especially as I knew there could be power lines anywhere along that flight path.

Tragically, the inevitable was about to happen! After a quick lunch beside the lake, we took off with each helicopter carrying troops for a 'dummy operation' along the Mozambique border. The idea was to distract the enemy from an intended cross-border operation that our security forces had planned for the next day.

Each helicopter pilot performed a short running take-off with his heavy load of troops, which was usually done when heavy-laden, and an unobstructed section was available. This placed less strain on the main rotor gearbox, and was an economical way to get airborne, where the pilot would run the aircraft on its nose wheel, to accelerate it through transition, before lifting off into the air.

Once airborne, each helicopter followed one behind the other across the lake, and the downwash from the rotor blades sent plumes of spray swirling off the surface as each machine built up enough speed to climb away. By the time I reached the other side of the lake, I noticed the other helicopters circling a point ahead of me. This drew my attention to the wreckage of the lead helicopter, which had just struck some electrical power lines and was now lying in a mangled heap. The crash killed everybody on board, including the pilot Paddy Bate and the detachment commander Bruce Snelgar, who I had previously met during my early days in the bush, and with whom we had just shared lunch.

The next day we entered Mozambique near a prominent feature called 'Monte Casino'. The invasion of Mozambique became even more confusing, as we had to fly in the thickest smoke haze I had ever seen. It was like flying in thick mist with visibility down to a few hundred meters, which made navigation extremely difficult.

An airborne JOC, which included Group Captain Norman Walsh, operated from a DC3, which flew along the border with Mozambique. I heard one message from this JOC when they sent a report of Russian tanks and a white combatant running from a contact.

When the ground forces encountered fierce resistance, they requested air support. Two Hunters were scrambled from Thornhill Air Force base and soon arrived overhead. One of the Hunter pilots, Sub Lt Brian Gordon, who was one of the top pilots on the first course I had instructed in Durban, fired a salvo of rockets. But he presumably misjudged his height under the horrendous conditions and was tragically killed when the pull-out was left far too late, and the aircraft smashed straight into the ground.

Later that day, a Canberra bomber flown by Kevin Pinkie, was also deployed to assist ground operations, but sadly, he flew into a range of hills after dropping his bombs. The thick haze would have contributed towards this, as he would usually have seen the mountain range in better visibility. But also, being a close friend of Paddy Bate, who had been killed in the accident the day before, many people thought that in his anger and grief, he might have also flown too low to press home the bombing run. Tragically, this Canberra accident added to the two other aircraft losses during the last two days.

During the operations, the JOC flying onboard the DC3, told me to land in the middle of the bush in Mozambique and switch off until they had another task for me. I distinctly recall the surreal feeling, that not far away, mayhem and carnage was taking place, and I was vulnerable to enemy forces, who could stumble upon our position at any time.

Later, the command DC3 instructed me to drop a 'stick' on the side of a mountain. But, without visual reference to the natural horizon, there was such poor visibility that I could not see that I was landing on a steep downslope. The main wheels touched first, and as the nose wheel was about to touch the ground, I had to pull back sharply on the cyclic to stop the helicopter from drifting forward. This caused one of the rotor blades to glance off the tail boom's top cover and damage a vertical fin. In conditions of better visibility, this would never have happened.

This call-up was the last time I flew helicopters on Rhodesia operations, but I already sensed that the war had escalated to a new and worrying intensity. I left with a heavy heart and a sinking realisation that we could no longer prevent the loss of this beautiful country. It was the place of my birth, which held so many great memories and a way of life dear to us.

In a few short months, the USA had twisted South Africa's arm so much, that they were forced to pull the plug on us in their bid to forestall their own inevitable outcome. Then, shortly after joining Court Helicopters in Cape Town, I remember the scorn I received from the dock operations manager, when he scoffed at my prediction, that international pressure would also force South Africa to the same forces of nationalism that we had tried so vainly to hold off, but within ten years, would be forced to succumb.

Then, almost ten years from the day that I made that forecast, South African Prime Minister JW de Klerk arranged for Nelson Mandela to be released from Robben Island.

Over forty years later, greed and corruption have resulted in Zimbabwe's collapse. Julius Kambarage Nyerere, the first president of Tanzania, told Mugabe to take care of "the Jewel of Africa", which had been Rhodesia.

23 ROBBEN ISLAND – CLOSE SHAVE

Sikorsky 58T servicing an oil tanker.

Roger Watt and I joined Court Helicopters together in 1979. However, our start in this company had not been well received. Going into a new environment can sometimes present problems. Still, I know it was exacerbated by Roger and my critical and judgmental attitude towards the standard of instruction. The training captains responsible for introducing us to ship-service operations on the Sikorsky S61 and 58T showed a lack of enthusiasm. Unsurprisingly, they did not take this too well to our mentioning this. Despite our assessment being well-founded from a military perspective, it did not help us win friends or influence people. As a result, it took years before the management considered that our previous experience or qualifications could benefit the company.

Unfortunately, one evening, at a cocktail function, we expressed our opinions to Richard O'Moloney, the Overseas Manager. He later brought this to the attention of the MD.

Of course, the two training captains took umbrage at our criticism and felt they should teach us a lesson in humility. When we underwent captaincy training, they ensured they gave no assistance or input. Whenever they flew with us, they declined to offer a single word of advice or constructive criticism. I found this a trying experience, especially as I already struggled with an identity crisis, and it did nothing to help my self-confidence as I battled to maintain a positive attitude.

For my benefit, I decided to compile an operation manual, like the one I had previously drawn up for No 7 Squadron, in the Rhodesian Air Force. After flying with several different captains, I realised the need for this as each one had a personal way of doing things, that made it difficult for the unfortunate co-pilots, who had to learn each captain's preference. Once more, I was not encouraged by these two training captains in my endeavour to standardise procedures. Eventually, Jeremy Labuschagne, the Operations Manager, took an interest. When the manual was finally accepted and approved, he rewarded my efforts by almost doubling my salary. This came at an opportune time, as our landlord had virtually doubled our rent. I immediately gave notice and signed a lease for a beautiful, thatched home with swimming pool, which would have been way beyond our means.

I eventually received my captaincy rating to carry out ship service operations on the Sikorsky S61, but only after appealing to the Operations Manager himself. My feelings of frustration became so unbearable that I felt if they held me back much longer, I would become so demoralised that I might never salvage my shattered self-belief. I asked for an appointment with Jeremy, who empathised with my frustration and asked when I thought I would be ready for a check ride. I told him I felt I was as prepared as I would ever be. He immediately arranged to accompany me the next day when I would be on duty.

My assignment that day was to re-supply a tanker as it steamed towards the twelve-mile position off Sea Point, before it turned north on its way to Europe. Jeremy joined me as the testing officer in the co-pilot's seat.

Jeremy set up the instrument flying screen to block my view outside the cockpit on that clear day. During all flight tests, the company used this system to simulate the procedures employed in fog, where we could still provide a service by descending to 50ft above the water, with the visibility reduced to half a mile.

During my flight test, we made a few trips to the vessel, where we took out one-ton loads of supplies, carried in under-slung cargo nets. Workers in the dockside hangar first packed the supplies into cargo nets on pallets, which they then brought out by forklift alongside the landing pad. After take-off, Jeremy hovered the helicopter above the load, which one of the hangar staff connected by a strop to the cargo hook beneath the fuselage. Once airborne, I had to direct the aircraft with the aid of the ship's beacon and radar. I also made radio calls to Cape Town ATC and Port Captain. After this, I contacted the ship's captain to establish their heading, relative wind across the deck, and what obstructions to avoid. Using this information, I could then plan the best direction to approach the vessel. Once I correctly identified the oil tanker on our radar, I established ourselves on the final approach, down to half a mile and 50ft above the water. At that point, I looked up and took over the flight controls. From that point, I had to decelerate the Sikorsky helicopter to a hover, position the load over the drop-off zone, and gently lower it onto the deck. The flight engineer would then operate the disconnect button to release the strop.

Once we had completed resupplying the ship, Jeremy instructed me to carry out a simulated diversion to DF Malan Airport, which was the standard procedure whenever the fog became too thick to make a safe landing at the dock's helicopter pad.

While we positioned for an ILS let-down onto runway Zero One at DF Malan, Jeremy hit me with all sorts of simulated emergencies, including an engine failure, which necessitated a single-engine approach. Then, just as I looked up to see the runway at 200 ft AGL, he announced a fire on the remaining engine. With an engine on fire, I would usually have to shut down that engine and pull its fire extinguisher handle. Fortunately, I had prepared myself well enough and elected to ignore the fire until I had made a safe landing. Once I overcame this last hurdle, he was satisfied that I was competent enough to qualify as a captain on the Sikorsky S61 and 58T.

Before this check, while still under captaincy training, I flew with a certain captain called Stan. Although he was not a training captain, the company allowed him to fly the helicopter in the co-pilot's seat. They also allowed me to make all the decisions and assume the captain's responsibilities from the proper captain's seat. In the early morning hours, we came on duty to do 'ship service' and take a single load to a vessel steaming up the West Coast, which was about an hour's flying time to the north of Cape Town.

We departed with the aircraft radar being unserviceable, as the met department had reported that the weather would be clear over the vessel and would fall within the minimum requirement for the task. However, once airborne, I noticed a bank of low cloud over the dock area, which extended to about 10 miles towards Robben Island. I decided when we returned to the docks with no radar, we might have to duck below this cloud to make a safe approach and landing.

Stan, like most of the Court Helicopter pilots, was an ex-South African Air Force pilot, and like many of us, had also flown the Alouette III helicopter. Rumours abounded about his strange behaviour, and most co-pilots tried to avoid flying with him. Having never been on a flight with him before, I had no first-hand knowledge of these problems.

We took off before 'first light', and on the way out, the initial part of the flight was routine, if not exactly relaxed. The early morning conditions were smooth and calm, with good visibility.

On our return, we homed towards the Robben Island NDB (Non-Directional Beacon) and once overhead the beacon, we commenced a 'let down' towards the pier by using a 'back bearing' of 179 degrees. I monitored the let-down, following company procedures, while Stan carried out the co-pilot's duties. He flew the helicopter on instruments and maintained a controlled rate of descent and a speed of 90 knots down to 1,500 feet above the sea, which he should have held until we reached five nautical miles from the dock area.

Then, under normal circumstances, we would have used the radar to home towards the pier and reduced the speed to 60 knots and the rate of descent to about 600 feet per minute. Then, as the aircraft got closer, the picture of the pier would become more distinct when the pilot set the radar to a bigger scale. To keep the aircraft descending towards the end of the pier, the captain would call out headings to steer and the descent rate to lose 300 feet for every nautical mile to maintain the correct descent path.

However, in this case, with the radar out of action, I could see at our present rate of descent, we would fly straight into the bank of low cloud. So, I suggested we abandon the let-down and make a visual approach by descending early. In this way, we could keep visual contact with the sea, as the cloud base was only a few hundred feet above the water. However, Stan ignored my suggestion, and radioed the docks operations staff, to get a visual weather report. They confirmed they could see Robben Island's lights and the beacon light at the end of the pier. Surprisingly, this information seemed to convince

Stan that if he stayed on the 179 degrees 'back bearing' from the Robben Island NDB, he would be able to descend safely through the layer of cloud.

I felt uncomfortable with this decision, but technically, as he was still the captain, he had ultimate authority over this flight. As we entered the cloud, he maintained 90 knots, although he should have reduced it to 60 knots at 1,500 feet. I brought this to his attention, but once again, he simply brushed aside my concerns. As we now had no reference to give us an actual distance to the pier, I started to get extremely worried and horrified by the predicament he had put us in and felt trapped and helpless to avert an impending disaster. Now, completely 'out of the loop', Stan remained oblivious to the danger to which he was blindly subjecting us.

At 500 feet, I started to pray for a break in the cloud below us and desperately searched out of the side cockpit window. Suddenly, a gap opened, and I was shocked to find we had overshot the dock area and were in a fast rate of descent over red rooftops on the side of Signal Hill. When I saw the houses, it confirmed my worst fears and gave me such a jolt that I snatched the controls out of his hands and pulled back hard on the cyclic. As the nose went up, I simultaneously yanked up on the collective lever and by taking this automatic and spontaneous action, it saved us from impacting the rising ground. After a few agonising seconds, we popped out through the top of the cloud with Table Mountain straight ahead.

Not saying a word, I turned 180 degrees, and once clear of the line of cloud, I set up a descent to duck beneath the base to make a visual approach and landing on the helicopter pad. Once safely on the ground, still shaken and angry, I handed the controls back to Stan while completing the 'shut down' checks. I expected some sort of apology from him for nearly killing us. But although he had sat back meekly throughout the unorthodox way, I had taken over control, I was stunned when his only remark was, "The ops staff were talking rubbish"!

Without making any further comment, I went directly to phone the Ship Service Manager and told him what had transpired and suggested he get another captain or co-pilot. He asked me to write a report, as no one else would commit themselves to paper. Jeremy Labuschagne, the Ops Manager, was contacted and decided to come in and fly the rest of the day with Stan.

That afternoon, I handed my report to him, with an extra copy for Stan. Later, Jeremy called me into the manager's office at the docks, where they were waiting for me. To kick off, Jeremy asked Stan to comment on the incident. But all Stan could do, was mutter something garbled about the back bearing from Robben Island and a cross-bearing of 330-degree radial from DF Malan. Sadly, this made absolutely no sense to either of us.

At that time, Court Helicopters had an offshore contract to support an oil rig off Rio Gallegos in Argentina and because of my report, Stan was taken off ship service and transferred to Argentina's bleak area, which earned him a lot more money. A few years later, he returned to take over a helicopter crop spraying operation that I had established. Unfortunately, while doing crop spraying, he lost his life by flying into electrical cables running across the farm. The helicopter was a complete write-off, and when they recovered his body from the wreckage, they found his pockets full of painkillers. On a subsequent post-mortem, X-rays showed that he had a brain tumour the size of a large coin, which went a long way to explain his irrational behaviour over the last few years.

As I entered our house, on the morning we nearly crashed into Signal Hill, before I had a chance to say anything about the incident, Tish asked me what had happened. When I asked her why, she described how, at four o'clock that morning, she had been woken by the bedroom door being opened and surprised when she saw a tall dark shape that looked like me, in the doorway, which said, "Hello, I'm dead!" To this, she responded, "Oh no you are not! I know who you are! You are the Father of Lies. I rebuke you in the name of Jesus!" And instantly, the shape disappeared.

I immediately realised, that this was the precise moment that the gap in the cloud had opened!

24 I THOUGHT THE FISHES SLEEP ON THE BOTTOM AT NIGHT.

I had been flying from Cape Town docks with Court Helicopters for a few months before my family left Rhodesia to join me. One Saturday, after being together for a month, I was on standby duty from 6 o'clock that evening. We had spent most of the day at home before I received a call to come in for a flight to deliver supplies to a ship south of Cape Point.

At that time, our duty roster entailed four standby duties during daylight, followed by four at night, and then four days off. The standby duties were carried out at home until the ops room called us to do ship service at any time of the day or night.

They would call us a few hours before the expected departure time for the first load, and we would sign on one hour before that. On most occasions, we would service the vessels as they steamed in towards Sea Point, where the positions they chose were either the twelve-mile, or seven-mile point. The closer the ship came, the less they would have to pay for the service. As the vessels reached their selected point, they would slow right down before turning away and continuing their journeys north or south of Cape Town.

On many days, no ships required our service, and I could spend the whole day at home. On other days, I would go in on my motorbike for a short trip off Sea Point, before returning home.

One day, just as I set off around the block for some fresh air, our neighbour arrived back home after taking his dog for a walk. Until then, I had never considered what he might have imagined and introduced myself. After a short chat, he said to me, "I don't mean to be inquisitive, but do you have a job?"

That Saturday afternoon, Ops called me to report for the co-pilot duty that evening on the Sikorsky 58T. The captain, Pete Woolcock, was an old friend of ours from the Rhodesian Air Force days and had already worked for Court Helicopters for a few years. Pete and his family had become our close friends, and our children had come to know him well. While I was attached to the SAAF and lived in Amanzimtoti, he had been the Ship Service Manager at Louis Botha Airport in Durban.

That evening, as I was getting ready to go to work, our two sons, Shane, who was fourteen, and Ryan, who was eight, asked if they could come to the docks with me. I realised that they were hoping for a ride in the helicopter, but I was not too enthusiastic about that, as I had enough on my plate with my co-pilot duties, without the added worry of their safety as well. So, I told them that I would take them if they did not bother Pete or ask him to come for a ride. I stressed that they would have to play around the hangar while I flew. Unfortunately, this failed to deter them, as enough was going on around the hangar to stop them from being bored. They had previously been fascinated by the hive of activity as tractors and forklifts buzzed around, taking pallets of provisions to the landing pad. These loads would later be carried in nets under the helicopter to one of the passing ships.

As we reached the Cape Town docks centre and started to walk across the helicopter landing pad, we saw Pete, who had arrived just a little before us. An effusive person, he greeted us in his hale and hearty manner and the first thing he said was, "Hello boys! Are you coming up for a ride with us?"

Now I had no option but to allow the boys to come along. As one of my duties, I had to brief them as 'passengers' on the flight's safety procedures and what actions they should take in an emergency. I fitted them with life jackets and explained what would happen if we needed to ditch at sea. I also told them that the helicopter would float for a while, and the flight engineer would launch an inflatable life raft we called a dinghy and help them climb into it. A lanyard attached the raft to a strong point in the cabin, which the onboard flight engineer would only cut loose when everyone had settled down when wearing fully inflated life jackets. I also showed Shane and Ryan how to pull the toggle to inflate the life jacket and stressed that they should only do this if the helicopter was floating in the water.

The Sikorsky 58T has a cabin that sits below the cockpit. The pilots would gain access to their section by climbing up the outside, using the foot-hold recesses and outside handles, whereas the passengers and flight engineers could climb straight into the cabin from ground level. Sometimes, the loaders would also put some cargo items in this compartment.

The engineer would wear a 'monkey chain' around his waist and attach the other end to a strong point in the cabin when we took underslung cargo. He would then lie on the cabin floor, lean out of the sliding door, and give directions to assist the pilot to position overhead the loaded net. Once over the load in a steady hover, a ground handler would attach the net's chain to the underbelly cargo hook.

The engineer would then give instructions to take up the slack and lift the load, which often weighed more than a ton. The helicopter would then be moved off through transition and accelerated into forward flight. They would safely set off into the wind to avoid obstacles and built-up areas whenever possible. The engineer would continue to monitor the load until the helicopter had accelerated to cruising speed and the cargo was stable and secure. He would then close the sliding door and sit back until the aircraft was on the vessel's final approach. At this point, he would give directions for the pilot to hover over the drop-off zone and lower the load onto the deck, before pressing the release button. The aircraft would then fly off for the next shipment or land on the deck to retrieve the neatly folded cargo net when possible.

On this flight, the vessel was unusually far out at about 70 nautical miles south of Cape Point. To make it worse, it was sailing in bad weather conditions. Violent thunderstorms had built up, and the met office told us that the wind over the vessel would be gusting up to 60 knots in heavy rain.

It was dark by the time we passed Cape Point and hit inclement weather. Deafening raindrops lashed against the helicopter, and we started to bounce and jolt around terrifyingly. This caused the blades to clatter in loud complaint at the unusual stress the turbulent condition was inflicting. I sat there, heart in my mouth, regretting the added responsibility of having our two sons sitting in the cold and dark cabin below. I thought of how Tish would feel if we lost our boys, or if she could ever forgive my taking them up in such stormy weather.

To add to my stress, the ADF needles, which we used to home onto the ship, were whizzing around crazily as they pointed towards every blinding flash of lightning that lit up around us.

With incredible difficulty, we finally managed to make radio contact with the vessel. When we spoke to the ship's captain, he gave us the ship's course and the relative wind across the deck and told us that the vessel was pitching by 13 degrees as it steamed into towering swells, with a solid spray of water crashing over the deck.

We finally picked up the ship on the radar and positioned the aircraft on the downwind leg, while descending to 500 feet for the inbound turn. I fought the controls to maintain as steady a speed as possible, but we had to increase the speed to 90 knots to progress towards the vessel. We switched on the windscreen wipers two miles from the ship. At half a mile and 50 feet above the thrashing sea, the blurred glow of the ship's lights showed up, but the landing light just reflected off the driving rain. So, we switched them off until later, to make seeing the ship's lights easier. We even had to ask for all the deck lights to be switched on. I sweated as I kept my eyes on the instrument panel, trying to adjust the heading and rate of descent, which Pete called out. He continued to give me directions until he felt it was safe to take over the controls and visually move the helicopter over its large painted 'H' landing area.

As Pete got over the deck, he had to pump the controls, while the flight engineer gave him directions to get over the right spot to let down the load. The windscreen wipers struggled to clear the deluge of water from the driving rain and spray that shot over the ship's bow. As the vessel pitched up and down in the twenty-foot swell, I called out the airspeed readings, which fluctuated between 55 and 70 knots, while Pete kept the aircraft in full forward flight to hold our position over the deck. He had to pump the collective control lever up and down to lower the helicopter and place the cargo net onto the deck before the flight engineer could release it. I was so tense throughout this that I could not wait to get away from the situation. I was also concerned about the threat posed by the ship's masts, as they swung and lurched up and down, only ten meters from the rotor blades. I gritted my teeth as we waited for the crew to pack up the net and attach it to the hook on the end of the winch cable. Under these conditions, I could hardly imagine how the deck crew must be feeling as they struggled in their yellow oil skin ponchos under the soaking spray to complete their task.

After an interminable period, we could finally move off and climb away to a more comfortable height above the sea. As we made our way back to base, the weather progressively cleared up. Amazingly, Cape Town's lights reflected off calm water, the cable station shone brightly, and the stars twinkled overhead.

When we finally completed the shutdown checks, I climbed down from the cockpit to meet the boys and took off their life jackets with shaking hands. As they thanked Pete for the ride, he replied, "That's OK, were you scared? Because your dad was!"

I signed off and thanked Pete for the flight but mentioned that I had regretted the boys coming along under those conditions and then drove off in our VW Passat station wagon to wind our way over the dock area's crisscrossing railway lines.

Shane sat next to me with little Ryan in the back seat as we drove away. We had not discussed the trip, and I tried to calm down after the hair-raising experience. As I was wondering how the boys felt about the flight, Ryan's sweet voice came from the back seat, "Dad, you know when we were flying in the rain? Well, I didn't know if we were in the water and if I should blow up my life jacket. But I tried to see if there were any fish first. And then I thought, 'But maybe fish sleep on the bottom at night."

Well, by the time he had finished telling us this, the relief of tension was so great that Shane and I started to roar with laughter and with the tears running down my cheeks, I could hardly drive the car.

25 BERNARD – LARGER THAN LIFE PILOT

I met Bernard after I had been flying for Court Helicopters in Cape Town for a few months. He was a good-looking young man with a square-faced Germanic look and about six feet tall with a good physique. He was also proudly fit, and passionate about many things; not the least of which, was his belief that he was duty-bound to promote the Afrikaans language. His parents had immigrated to South Africa from the Netherlands, and he had joined the *'Oswa Weerstand Beweeging'*, a youth organisation like the 'Hitler Youth Movement'. I presume this is how he became imbued with zeal for the Afrikaans culture, language, and traditions.

Although English is recognised and accepted as the common language in aviation worldwide, during Apartheid in South Africa, the policy was for air traffic controllers to be equally conversant in both official languages. However, notwithstanding the flight safety considerations with foreign carriers that flew in and out of DF Malan Airport in Cape Town, Bernard insisted on making all his radio transmissions in Afrikaans, despite being fluent in both languages.

Many of the pilots who flew for Court Helicopters were ex-Rhodesian Air Force pilots of British descent, and with characteristic arrogance, they had total disdain for the Afrikaans language. However, like many other Afrikaners, Bernard was still caught up with the Boer War's atrocities nearly a century ago, and the British were still very much the enemy. These pilots from Rhodesia were a thorn in Bernard's flesh, especially when they pronounced his name, *"Bernard"*, and did not roll the 'R' and neglected to emphasise the 'A' sound at the end. Since I enjoyed learning other languages and was happy to speak to him in Afrikaans, I had the dubious privilege of having a window into his life.

At that time, he was a co-pilot on the Sikorsky 58T helicopter, which operated out of the Cape Town Dock heliport, from where we resupplied ships rounding the Cape of Good Hope. As far as I remember, he had started as an aircraft technician and had been allowed to undergo pilot training with the Company. I met him when he returned from a stint in Rio Gallegos in the south of Argentina, where Court Helicopters had a contract to resupply offshore oil rigs.

Bernard was still single, but desperately longed to have a meaningful relationship with someone from the fairer sex. Unfortunately, he found it difficult to meet anyone who could live up to his high standards, as most of the ones he met were too stupid, too untidy, too lazy, or all three at the same time. Also, no prospective young lady could cook better, sew better or iron better than he could.

He was also intolerant of what he deemed to be slipshod behaviour. However, to the young ladies at the church where he worshipped, he was an eligible bachelor who had made good money on overseas contracts. However, none of them could meet all the required attributes, plus be endowed with great beauty, and it was either his way or the wrong way.

Bernard was also tremendously fit from powering his racing bike long distances around the beautiful Cape Peninsula and he would often regale me with hilarious stories, which were profoundly serious to him.

A pet gripe of his was his penchant for keeping the City of Cape Town and its environs trash free. This became a serious matter to him, and he would never hesitate to reprimand anyone who littered. One day, while he was cycling, a motorist threw a banana skin out of his car window. This immediately affronted Bernard's sense of decency. He was so indignant that he snatched up the unwanted debris and took off after the offender. Catching up with him at a traffic light, he hurled the offensive litter at the surprised driver through his open window!

Later, he related another cycling incident to me. This concerned a yapping dog, with which he had experienced continual problems, as it would always chase after him, and snap at his heels whenever he passed a certain spot. One day, as Bernard was kicking out at it, he lost his balance and came crashing off his bike. Picking himself up, he furiously abandoned his racing bike, chased after the culprit, followed it through a garden gate, an open front door, and straight into the lounge of the bewildered owner, who was peacefully reading a newspaper. Bernard grabbed his quivering dog by the scruff of its neck, and in front of his equally quivering owner, gave it a hefty wallop.

Stories also abounded about Bernard's parsimony, when he had been on contracts in Argentina. The company paid crews a handsome daily allowance, and it was Bernard's primary intention to accumulate as much money as possible while he was there.

The company's personnel lived in a rented house, and usually shared catering expenses. However, Bernard brought food from South Africa for his exclusive use. His stash included, eggs, a large jar of peanut butter, and Lyle's Golden Syrup, and he became furious if other crew members helped themselves to his private supply. So, he pencilled his name on his provisions and attached strings to each egg, which he then led down the passage to tie to his big toe when he went to bed.

The others would hotly deny his accusations, until one day he returned unexpectantly and caught one of them red-handed. The guilty party had just used two fingers to scoop a dollop of peanut butter out of the jar, which he was about to put in his mouth. But with a few giant steps, Bernard bounded across the intervening space, grabbed hold of the offending thief's wrist, and stopped him from eating it by twisting the peanut butter-laden fingers into his own mouth.

Bernard's one pride and joy was his motorbike, which he kept in mint condition but periodically enjoyed taking exhilarating bursts at excessively high speeds. Unfortunately for him, on one of his return trips from Argentina, he was caught in a speed trap by a couple of Afrikaans police officers who tried to speak to him in both Afrikaans and English. To avoid prosecution, Bernard pretended he could only speak Spanish, which to his credit, he had mastered during his Argentinian contracts. Then in total frustration, the one policeman said to the other, *"Los die poephol!"* (Let the a---hole go), which was poetic justice for someone who would never have tolerated such a derogatory insult. However, it was poetic justice, especially as they said it in Afrikaans.

Sometime later, Bernard cycled into Cape Town. As he emerged from a business complex and was putting on his crash helmet, one of these policemen recognised him, but Bernard took off before they had a chance to apprehend him. By the time the policeman reached his motorbike, Bernard had raced away with the policeman in hot pursuit. With Bernard's small lead ahead, he just had enough time to turn into a crossroad then immediately ducked into an open driveway. Still sitting in the saddle and with no time to unhook his shoes from the pedals, he simply fell sideways behind a low hedge. By lying in this position, he succeeded to elude being caught and just listened to the policeman's motorbike tearing up and down the road, in search of him.

One day in the ops room, I was quietly amused when one of the John Rolfe Air Sea Rescue team innocently related a story about Bernard's brilliance as a pilot to Jeremy, the Ops Manager. Jeremy's eyebrows rose higher and higher in despair, as the storyteller went on to describe how talented Bernard was at controlling the helicopter 'cyclic' with his knees while eating his sandwiches. But this only confirmed that when it came to his flying, he was a bit of a loose cannon.

One beautiful Sunday, we flew together on a sortie in

a Sikorsky 58T. On returning on the last leg from servicing a ship off Cape Point, Bernard first asked if we could look at a large metal beam lying on some rocks off Chapman's Peak. The metal beam was lying across some rocks on the Hout Bay harbour shoreline, which he felt he could salvage for scrap metal. Once we had a look at the length of metal, he asked if we could do a flypast along Sandy Bay Beach. Hidden by thick impenetrable Port Jackson bushes and huge sand dunes, Sandy Bay was an illegal but secluded nudist beach which could only be accessed by a long rocky path from Llandudno. Unfortunately, there happened to be a Cape Times photographer on the beach as we flashed past. He just happened to be standing in a strategic spot to capture the helicopter as it flew past at 50foot and about 200 meters out to sea. Although just a few nudists were casually strolling along the beach with their heads turned towards us, he published the picture in the Cape Times the next day. The Sunday paper sensationalised the incident, with a caption that read, "Court Helicopter pilot sends nudists running for cover from stinging sand, that was whipped up by the helicopter blades". As I was the captain of that flight, the operations officer summoned me to the head office to explain the embarrassing publicity. As a punishment for my indiscretion, I lost six months seniority.

A few years later, Court Helicopters had a contract to operate a Sikorsky out of the Anglo-American base at Kleinsie on the West Coast. The crews all lived in caravans with an enclosed lawn between them. Apart from this, we shared a brick-built kitchen and living area. One day, while the other pilots and engineers were away on a trip, Bernard and I started chatting. I made the mistake of relating a story from my time in the Caribbean, when I had arm-wrestled an aircraft technician called Johnny, an American rough diamond, with whom I had worked on a contract there. Bernard then went on to tell me how he had learnt a trick at school whenever he got into a fight. He explained how he would wrestle his opponent in a neck lock to the ground. Then, without any warning, he leapt on me, threw me onto the lawn, and viciously throttled me around the neck, in a vice-like grip. His extraordinary strength made it difficult for me to breathe, and I tried to stay calm while he kept repeating, "Try and get free. Try and get free!" With panic setting in and feeling completely helpless, I tried to appease him as I gasped out, "OK Bernard, you have proved your point."

I am not sure how this would have turned out, as he did not seem inclined to let up, but, fortunately for me, just at the point when I knew it had gone too far, the other crew members returned. I felt rather foolish, but it was still with great relief that I could struggle back to my feet. It must have looked strange to see two pilots grappling on the ground but at least it prompted Bernard to let go and portray the complete picture of innocence.

After writing this account, I spoke to Lawrie Aenmey, a good friend from Court Helicopters. I learnt that he and Kathy had retired near George in the Eastern Cape, and a helicopter company was employing him as one of the temporary pilots flying out to a SASOL rig off Mossel Bay. He mentioned that the same company also used Bernard. Lawrie went on to say that Bernard was still the same and, at that time, was in his fifties and still unmarried.

I then sent Lawrie a draft copy of this story. When he told Bernard that I had written about him, he seemed proud of the fact.

In my first book, being mindful of his strength and unpredictable nature, I took the precaution to change his name in this narrative. But since then, Bernard and I have re-established contact with the aid of Facebook, and I felt brave enough to send him a copy of this account. I then waited with bated breath for his response but was relieved when he replied to me in Afrikaans and told me that he had not laughed so much in a long time. However, he refuted the story about his parsimony in Argentina and insisted that it related to another pilot there. So, as it is hearsay anyway, I cannot vouch for its validity.

26 RUSSIAN APHID – MIRACLE

When I left the Rhodesian Air Force to go crop spraying, I was on the crest of a wave in my career. I had just spent the best two years of my life on a top-secret assignment, where I had ostensibly joined the South African Air Force as a Major, during this time. But, despite our all wearing SAAF uniforms, it was still the AFS (Advanced Flying School) part of our Training Squadron, that now operated from the base at Louis Botha Airport to the south of Durban.

However, living in the holiday resort town of Amanzimtoti had shown me a glimpse of an exciting world outside the Air Force, which made me yearn for an opportunity to have a taste of it. But barely a month after leaving and joining a Rhodesian crop spraying firm, I survived a severe helicopter accident. Following this, the novelty quickly wore off, and in no time at all, the move filled me with regret. That first season of spraying subjected me to abject misery. Once again, I had made a spontaneous move, which I had based on irrational thinking.

I had left the Rhodesian Air Force, just when it seemed I was about to be destined for better things. I had qualified as a flying instructor on both fixed-wing and helicopters, and become an air weapons instructor and instrument rating examiner. Now, I sweated, hour after hour, day after day in a hot Perspex bubble, while I buzzed a few meters above the ground to spray cotton fields. There was hardly a word of appreciation from the farmers, who would watch while leaning on their Land Rovers. They would then disappear for tea or lunch, and later return to stand by idly, to keep an eye on things. They appeared mesmerised by the spectacle, while forgetting that the pilot was not just a robot or part of the helicopter itself. On one sweltering day, I was so thirsty that I sent a handwritten note to the farmer, to ask him for a cup of water. To his credit, he was so abashed that he brought me a long cold drink in a glass.

Once more, my natural flying ability pulled me through this unhappy time. Despite being in a constant state of depression, I reached a competent standard. I achieved the same results as the Ops Manager, Dave Thorn, who had previously flown in the Fleet Air Arm as a helicopter pilot. I had met him before when he had joined No 7 Squadron for a few years, when I was a flight commander and instructor there. At that time, I remembered his approach to flying was a bit unconventional and he came across as somewhat of a maverick in the Air Force. But when I joined Agricair, I learned many valuable tips from him.

By this time, some of my friends had moved to Cape Town, where they were now worked for Court Helicopters. Here they were busy flying Sikorsky Helicopters to supply oil tankers and other vessels on their way around the Cape Peninsular. I decided this would be right up my street, so after that, as I skimmed up and down the cotton fields, leaving a swathe of white insecticide swirling into the crop, I visualised myself flying over the cold, the indigo-blue sea off Cape Town. I acknowledged that I had made a wrong move and resolved to get out of this job as soon as possible.

When my contract with Agricair ended, I joined Court Helicopters with Roger Watt, one of my ex-Air Force friends. After validating our Rhodesian Commercial Pilots Licences, we underwent training on the Sikorski helicopters. Once qualified, the company transferred Roger and me to Durban. But after a short spell, I was sent back to Cape Town.

I had still not regained the self-confidence which had taken such a knock when I left the Air Force, and I struggled to find my sense of identity in this new environment. In retrospect, I am grateful that I had to face this problem early in life. Since then, I have learnt that when men retire after dedicating their whole life to one career, they grapple with a similar identity question, which has become intertwined with their work status. I have also read that men's average life expectancy after retirement is between two and four years.

In time, life in Cape Town settled into a comfortable routine. But although Cape Town is beautiful, it did not feel like part of the African continent. We also struggled with the low salary, which did not allow the family to enjoy the standard of life we had taken for granted in Rhodesia. We had no option but to pull in our belts. Whereas our meals had previously included high-quality meat at a reasonable price, it now became a real luxury, and we often ate a soya substitute.

Apart from these deprivations, we enjoyed a relaxed lifestyle and relished many off days and time spent at home between flights. I later discovered that neighbours thought I was unemployed, as they were unaware that I would go to work for only a few hours, at all times of the day and night.

It took a few years before I could believe that I was no longer an Air Force pilot. Nevertheless, we made many new friends through the Pinelands and Wynberg Presbyterian Churches. Our three children also made friends at school, and Tish took advantage of the quick and easy access to the seaside and other scenic picnic spots.

Jeremy Labuschagne was the Operations Manager when I joined Court Helicopters. An ex-SAAF helicopter instructor, he was a sophisticated operator in every respect. Being overly ambitious, he eventually rose to the position of being the company's MD. He would sometimes help at the docks on ship service operations in those early days. As a designated training captain, he was also available to fly as a co-pilot, and after I had qualified to fly as captain, he would occasionally fly as my co-pilot during nighttime shifts. He would keep very much to himself when not flying, which contrasted with the other pilots and flight engineers, who would sit chatting in the crew room while drinking copious cups of coffee. Things seemed to weigh heavily on Jeremy's mind, and whenever we had to wait around at the hangar between flights, he would be seen in deep thought while pacing the hangar floor.

One night, as we flew together, I was surprised when he started to chat. He never made idle conversation, but when we were on our way back from servicing a vessel, he asked out of the blue, "George, you've done crop spraying haven't you?" Completely taken off guard, I was hesitant and a bit suspicious as I answered, "Y-es?" remembering how unhappy I had been at that time. Undeterred, he continued, "Because I was thinking of introducing a helicopter for herbicide application on the wheat crop in the Western Cape. Maybe you could give us some pointers?" I was so taken aback by this unexpected discussion that I was at a loss for words.

After we landed, I went upstairs to the hangar crew room above the ops room. Jeremy, in the meantime, busied himself in some other area. As the next vessel was due only two hours later, it was not worth going home. But while I sat alone in the crew room, Jeremy's words kept resonating in my head, and I could not repress a feeling of excitement that was bubbling up in my spirit. I was aware that it was planted from above, as I would not usually have shown the slightest interest in anything to do with crop spraying. But the more I tried to suppress this bubble of excitement, the more it persisted.

When I finally got home, I tried unsuccessfully to get some sleep, but I could not silence the stirring in my spirit. After this restless night, I knew I had to call Jeremy at his office the following day. When his secretary put me through to him, I asked if I could come in and see him sometime. So, he invited me to come in straight away.

On the drive to the head office at DF Malan Airport,

the excitement continued to well up inside me. The moment I sat down in front of him, I blurted out, "Jeremy, I've been thinking about our discussion last night. If you want, I am willing to set up a crop spraying operation for you". I then stipulated though, that I would first need to visit Harare to talk to Mike Saunders at Agricair. While there, I could obtain the design specifications for the ground equipment and glean as much information about their *modus operandi* as possible. I would like to learn how they set up the spacing and calibrated the spray nozzles for different spray requirements. Although Jeremy always maintained an air of aloof charm, his response was immediately enthusiastic as he asked, "When is your next off day?" "Tomorrow," I replied. "OK, just hang on a second," and he buzzed the intercom on his desk to summon his secretary. When she walked in, he instructed her, "Gillian please book Captain Wrigley on an SAA flight to Harare tomorrow." By the time I had finished another cup of tea, she was back in the office to say, I could collect my tickets from the terminal.

As I left head office, my head was spinning at the pace that things were going. We contacted Tish's parents and asked them to collect me from Harare airport the next evening. I also asked if I could borrow their car to go to Charles Prince Airport for the next couple of days. So naïve, it never entered my head to ask to use the Company's expense account to hire a car or book into a hotel. On my return, I never even considered to submit an expense claim for other allowances. Of course, Jeremy was not going to enlighten me either!

It was also strange when I returned to Harare, so much sooner than I would ever have imagined. Now, the country had a black prime minister, but on the surface, nothing seemed to have changed very mush.

After I had been collected from the airport, and we were travelling the back road to the Ruwa farm where Tish had grown up, I could see elephant grass standing ten feet high on each side of the road. As I sat looking out of the car window, I became so choked up with emotion that I could hardly breathe. It was all so beautiful and familiar, that it made me feel like a spirit who had returned from the dead. I did not feel part of life, but more as though I was able to see everything around me from the outside looking in. This feeling remained with me for the duration of my stay in Zimbabwe, as I continually caught glimpses of places that evoked floods of memories,

Considering I had left Agricair after just one season, Mike Saunders, the other pilots, and the staff treated me with unexpected kindness and were happy to furnish all the information I needed. They even gave me one of the contract forms, which they got farmers to sign before spraying. By the time I left them, I had a good understanding of how to calibrate and arrange the nozzles on the spray boom and how the pilots' bonus structure worked.

When I returned to Cape Town, I contacted the regional manager of the *Suidweslike Kooperasie*, the farmer's co-op for the wheat-growing area of the Western Cape. I found out as much as I could to ascertain the number of hectares that might be available. Most of the wheat growing farmers had been using fixed-wing aircraft for several years now, after different helicopter spraying companies had left them in the lurch in the past.

While the hangar staff set about manufacturing a mixing tank and trailer, I asked Peter Hill, an accountant friend from the church, to help me work out a bonus scheme that would earn me four thousand rand if I flew the maximum permissible hours in one month. But at that time, my salary was only nine hundred and twenty-five rand. I estimated how many hectares I could spray for each flying hour while also considering moving between farms.

Peter drew up such a professional presentation that Jeremy was full of praise. However, when he took the report to the MD, Aat Slot, who investors had recruited from Holland, he flew into immediate outrage by the commission structure. Especially when he realised, that if I managed to fly one hundred hours each month, I could earn more than him. His manner and attitude became so ugly, that it put my back up, and I responded, "In that case, we should just drop this whole thing." To his credit, Jeremy managed to calm us both down with his inimitable smooth manner. With a slight gesture towards me, he assured the irate MD, that it would be impossible for me to keep that busy for a whole month, or even achieve the anticipated hectares. Furthermore, it would be a miracle to attract that much business in the first season. So, by the end of a highly charged meeting, the report received a begrudged blessing of approval. We then went to the financial manager Karl Pitman, who also baulked at the figures, but Jeremy managed to smooth over his misgivings as well. He then arranged for me to submit a monthly claim against my returns, which would show the number of hectares I had sprayed that month.

While the hangar workers at DF Malan manufactured the ground equipment, we had the booms fitted to an immaculate new Bell 47G helicopter. We then attached the recommended size and number of nozzles for herbicide (weed killer) application. With a bit of blue dye added to the water in the hopper, I carried out numerous trials by flying at 10 feet AGL and 60 knots, while spraying onto sheets of white paper. Then, after each run, we would measure the spread and size of the droplets. The head office hangar technicians took on the project with great enthusiasm and seemed to enjoy getting things up and running while spending time outdoors. However, the one glitch that bugged us, and stayed for a long time, was the lingering problem of trying to start the helicopter on cold mornings. It frustrated and baffled the engineering staff, who changed magnetos, fuel lines, fuel and air filters, and fuel pumps to no avail. All along, the problem put my patience to the test. While I sat in the cockpit for hours on end, I would push the start button in short bursts to avoid burning out the starter. By the end of the day, this would make my thumb ache.

While we still struggled with this problem on cold winter mornings, we lost valuable time when I should have been busy spraying numerous hectares of beautiful wheat land under ideal conditions. After considerable frustration, the technicians eventually solved the problem. But before that, they had swapped so many components that I could not recall what they finally did to rectify the situation. Like many mechanical problems, I think it was probably something quite simple.

The other difficulty I experienced, was to produce the contract book, which each farmer had to sign before I would start spraying their fields. As all the farmers were Afrikaans speaking, I needed to translate it from the one I had drawn up in English. I was anxious to have the duplicate books finished and printed when ready to commence any spraying operations. The company asked me to give this job to an arrogant lawyer. But he dragged his heels to such an extent that in desperation, I had to ask someone from an Afrikaans newspaper to help with this translation. True to form, when the lax lawyer heard that I had bypassed him and sent everything off to the printers, he hurriedly rushed off a slapdash copy to us. Then, to add insult to injury, he had the cheek to send us a fat invoice for his shoddy work!

It took longer than I had hoped before we were ready to start spraying. But then, clients were not exactly lining up either. But one day, out of the blue, we received an urgent call from a Montagu's farmer in the Klein Karoo. He reported an outbreak of *"Rusiesekoornleis"* (Russian Aphid), which suddenly catapulted us into the deep end.

Up to then, we had been calibrating to apply herbicide

onto the wheat crops, but this was a completely different ball game. This first request needed us to spray insecticide instead, which had to be delivered like a fine mist onto the crop in tiny, closely spaced droplets. This meant we needed to switch to entirely different apertures and spacing between the nozzles. With hasty preparations and a lot of guesswork, we were quickly on our way.

The company allocated Hannes Eksteen to the operation. He was a good-looking young Afrikaans speaking engineer. Sent to assist was an amiable and intelligent Cape Malay, Edwin, whose ancestors had been brought in the eighteenth century as enslaved people from Southeast Asia.

The day we had to position for the first job, there was low stratus cloud over a large area of the Western Cape, which extended right into the mountain range to the northeast of Cape Town. I suggested that the support truck go ahead of the helicopter and wait for me just off the road once it emerged on the other side of Du Toit's Kloof. However, when I reached the top of the pass where the road crests the first ridge, I encountered low clouds touching the side of the mountains. But by keeping visual with the ground, I managed to squeeze through a small gap over a saddle, just beyond the exit of the newly built Du Toit's Kloof tunnel. The road then dropped steeply between the valley's impressive vertical walls as it zig-zagged towards Worcester. Although managing to sneak in, I could find no way out at the other end. So, without an exit from the valley, I decided to land just off the road, in a small hotel's car park, where the manager kindly allowed me to wait it out until the cloud lifted enough to continue again. When I finally managed to make it through the pass, just before Rawsonville, I saw the bakkie (truck) waiting there. Hannes and Edwin were surprised to see me. When they had driven through the pass, they had not thought it possible for me to make it through the thick mist in this little Bell 47G helicopter, which had no fitted flying instruments.

On arrival at the assigned farm, I was surprised to see how small their fields were. I could also immediately see that the crops looked noticeably stressed, not high enough and turning purple. The farmers had tucked most of the cultivated fields between narrow valleys in this mountainous terrain.

While we set ourselves up alongside the first field, I noticed many other farmers hanging about to watch proceedings. In this area, there were predominantly Afrikaans-speaking folk, who still held prejudice against using an English-speaking pilot. However, Hannes, the handsome blue-eyed Afrikaans engineer, excited some interest among the hopeful mothers, which made it an easier pill for them to swallow.

As an intrinsic exhibitionist, I enjoyed the chance to show off and provide some entertainment. I buzzed up and down, skimmed under electricity wires, and executed steep, heart-stopping torque turns at each end of the field. Once I completed the first load, I screamed into a nose-high quick stop and plonked my craft down beside the mixing tank. While Edwin pumped the following mixture into the hopper, Hannes attended to a few leaking nozzles. As I looked across, I saw a group of spectators wander into the crop to inspect the effectiveness of the insecticide application. Although I thought it might be a bit premature, I quietly prayed they would find some dead aphids lying there.

Once the farmer had signed the invoice to confirm I had successfully sprayed his lands, the onlookers approached me, one after the other, to book appointments for the rest of the week.

But before I could continue with these farmers, I had to attend to a previously booked farm on the outskirts of Durbanville, not far from Cape Town.

I arrived there the next day. But I was surprised to see the wheat lands spread across such an undulating and contoured area, where the fields I had previously sprayed in Rhodesia had been over almost flat ground.

This farm was a real eye-opener that proved to be a nightmare in many ways. For a start, the farm manager who met me, started to speak to me in Afrikaans, saying, "I can speak English but feel that the English people should learn to speak Afrikaans!" He then continued in Afrikaans to describe the complicated layout of the lands. Fortunately, I had taken Afrikaans as my second language at school. But apart from one oral test, I had never actually learnt to speak it. I had also never sprayed lands that disappeared over hilltops to plunge steeply down the other side. The shapes of the fields were also incredibly complex, without a single straight line to use as a reference.

The farmer employed coloured workers who were supposed to help me keep straight lines. However, the steeply undulating terrain made this almost impossible to achieve. The markers, who worked in pairs and had to do the job without any protective clothing, held a twenty-metre rope between them. The farmer had placed each pair of markers on the downwind side of the first field and instructed them to move towards the other side each time I flew between them.

It would have been academic if any field had a single straight edge. None of the markers could even see any of the others, which made the line higgledy-piggledy before I even began.

When I sprayed between a pair, one man was supposed to stand still while the other crossed over at ninety degrees to my line of spray. Unfortunately, they were also terrified by this crazy metal flying insect that buzzed between them. Instead of moving at ninety degrees to my flight path, they would just head off towards the nearest boundary on the other side of the field. Apart from this, I could not see more than one pair at a time. Then, as I passed over a high point, the following markers would never be where I expected them to be. All I could do was hold the same direction but struggled to maintain a constant airspeed. I would continually stagger up one side, and then plummet down the next. Whether I sprayed between the markers or not, they would simply move if I passed anywhere near them. In the end, my only hope was to do each field by eye and pray that the wind was strong enough to spread the insecticide through any missed sections.

The other thing that proved to be a headache, was that the farmer thought he could kill two birds with one stone by adding crushed seaweed to the insecticide. This made the mixture too fibrous and continually caused the nozzles to clog up. Each time I landed for the next load, I had to direct Hannes to the blocked nozzles. He would then attempt to blow each one out or clear it with bits of grass. However, I shuddered to think what effect the insecticide was having on him!

By the end of the day, I felt ill, feeling that the farmer would sue the company for my shoddy work. When I returned home for the weekend, I told Tish how worried I was, as this was only the second farm I had sprayed. The following day, we decided to drive out to the farm to inspect the lands. When we got there, Tish sat nervously in the car, while I tromped amongst the wheat to look for live aphids on the plants. The next thing she saw was me running down the hill, leaping in the air with arms spread in praise to our merciful God. Wherever I looked, I saw dead aphids lying everywhere, which reinforced my conviction that the Lord's hand was on this project. He had obviously, been fully aware of the impending outbreak of Russian aphid and put it into my lap.

The rest of the season was a marvellous experience as I stayed in the cockpit from the first light of the bitterly cold winter mornings to dusk in the evenings. I sat and ate sandwiches at lunchtime while Edwin pumped the loads into the hopper. I washed these sandwiches down with cups of strong black coffee, brewed by the farmers over burning coals in salty water and without sugar. The only opportunity I took to stretch my legs was when I had to shut down to refuel.

It was still in the *Apartheid* era, and the country did not classify Edwin as 'white'. So, when we checked into the small country hotels, they always wanted to put him into substandard accommodation at the back of the hotel. However, it created quite a stir, when we insisted that he stay in the same rooms they offered to Hannes and me. Inevitably, the managers only agreed if we guaranteed he did not upset the other guests by using the bar and dining rooms. Unfortunately, when some of the farmers accommodated us, they would sometimes put him up in a barn. But one morning, it mortified me to learn that he had chosen to sleep in the vehicle instead. It was impossible to insist that farmers should offer Edwin the same treatment they gave us, as they were under no obligation to provide any accommodation at all.

Flight and duty time was supposed to commence when the engines started. But I asked the engineers to fit a micro-switch to the collective, so it would only activate the hour meter when I raised the collective to get airborne. In this way, although I sat in the cockpit for nearly 12 hours a day, I was able to stay within the 8-hour legal limit.

There were also fantastic moments when *en route* to the next farm, I could take a break and land on top of a snow-covered mountain. Doing this allowed the bakkie a chance to get set up and mix the next load before I arrived. During these special times of escape, I soaked up the isolation and beauty up there, and on one of these occasions, I collected some mountaintop snow, which I put in my thermos flask to take home to show the children. I dread to think how I would explain to the company why I had landed on the mountain if I had not been able to restart the helicopter.

During that season, we were in constant demand and called to difficult areas nestled in narrow valleys in the Klein

Karoo mountains. The one unforgettable perfect day was on a Ceres Valley farm, with ideal weather for spraying in this famous fruit-growing area of South Africa.

The farm was west of the Matroosberg Range, opposite the Hex River Valley's beautiful vineyards. A gentle breeze blew across the length of this broad field on a 300-hectare block of wheat, with long and flat spray runs without any hazards.

At lunchtime, each time I came to land to take on the next load, the farmer brought something to eat while I sat in the cockpit. On the first stop, he offered me a freshly cooked sweetcorn on a skewer. The next time, he gave me a delicious juicy lamb chop, which I ate in my hands. I had never tasted anything quite so good. So, when he got me another one, I just had to shut the engine down to savour the moment. It was such a glorious day, with clear blue skies, that while I relished this succulent chop, I sat gazing at the crisp white snow shrouded, Matroosberg range. Then to top it off, the farmer brought a large mug of fruit salad and cream! By the time I was ready to continue, I felt nourished and refreshed in both body and soul.

As though this day had not been perfect enough, when I asked the farmer to sign off and thanked him profusely, he asked which I preferred, wine or brandy. Somewhat surprised, I blurted out, *"Brandewyn"*, and as a token of his appreciation, he presented me with a bottle of 10-year-old KWV export brandy.

Although I had not been happy while crop spraying in Rhodesia, this was altogether another kettle of fish, and I enjoyed everything about it. The farmers were so appreciative, that they continually offered us warm-hearted hospitality. The weather was also invigorating, and the snow-capped mountains were exhilarating. For the next three months, I flew six days a week, but we had to take a compulsory break each Sunday in the strictly religious Afrikaans farming community. We would generally leave the helicopter at the farm we had been spraying. Then, we would drive back to Cape Town on Saturday afternoon to enjoy time with our families. During the drive home with Edwin at the wheel, I was able to unwind, in an atmosphere filled with light-hearted fun, when we would joke and relive recent and humorous incidents.

Towards the end of the season, on one Friday, a farmer called us to an area north of Cape Town. This was close to the West Coast, and to the edge of the Western Cape's wheat-growing area. When we arrived at the farm, a young Afrikaans man in his twenties met us. He was disorganised, with only two farm workers to assist him. When I inspected his crop, I could see that it was a sad sight. The farmer had planted the wheat, spread out on small fields, where I could immediately see, by the purple discolouration of the stunted plants, that it had been badly infested by Russian Aphids.

The farmer invited us to eat first, and I was shocked when I saw the shack, he called home. He first offered us a bowl of shredded venison, which we had to eat in our hands, as it seemed that he did not use knives and forks.

Water was also quite a problem for him, as he had to bring in 200-litre drums from a borehole to the small pockets of wheat fields, scattered all over the farm. The whole exercise was very time-consuming, as we had to keep moving from one area to the next. As it became apparent that we would not finish the job by the last light, I suggested we stop and leave the helicopter overnight and return the next day. But he pleaded with me to complete the job, as he was desperate to get to a wedding in Cape Town the next day.

So, against my better judgement, I continued spraying. Eventually, it became so dark that I could not see the RPM gauge and the only way I could control the revs was by listening to the engine's pitch. I continued to fly until I could not even see the two workers, who would help me line up for the next run. Then I noticed a tiny flicker of light that indicated the markers' positions. These lights gave me something to fly towards, and I only later discovered that they had been striking matches!

When I finished the load, I had no idea which direction to fly to get back to the mixing tank, until Hannes, realising the problem, lit the bakkie's headlights, which provided enough light for me to see the ground for a landing.

Unfortunately, there was still one more load to uplift. As the adrenalin surged through my veins, I cursed both the young farmer for pressurising me, and myself, for getting into such a precarious predicament, as I allowed the crew to pump in this last load and took off once again! By now, it was so dark, that I could not even see the ground, and the only external reference to my height was the lit matches at each end of the field. So, I made sure I erred on the safe side and flew about 30 feet up. As there was no wind, I felt confident that the spray would sink onto the crop, but I continued to curse my stupidity as I flew.

However, because there was no wind to drift the chemical away from the next run, the swathe from my previous run started to smear the windscreen. This made it even more challenging to see anything at all. To add to this, each time I flew through the lingering spray without a mask, I had to breathe in the poisonous insecticide.

I was relieved to make it back to the vehicle at the end of the load, where we left the helicopter until we could return the next day. As we drove away, I was stunned by what I had just done, and wondered if I finally needed to have my head read. My chest was also burning from breathing in the insecticide, and when I got home that night, I was shocked to cough up dreadful green-coloured phlegm.

Compared to the previous risks in my flying career, this

must surely rank as the most stupid!

There were many wonderful and frustrating times along the way. But I still accomplished more than my most optimistic expectations. By changing the calibration to deliver the chemicals at an increased flying speed while spraying, each month I managed to make the money I had hoped for. I also scheduled the program to spend the least time flying from one farm to the next. It was an unusual season, with excessive demands on helicopters being called into tricky areas, tucked into small valleys, or covered by a network of overhead electrical cables, which precluded the use of fixed-wing aircraft. Many farmers, who booked our service that season, had never previously considered aerial spraying.

There is a saying in Afrikaans which goes, "*Een man se dood is n ander man se brood*" (One man's death is another man's bread) which was the case this time. Court Helicopters and I just happened to be at the right place at the right time, as we did well out of the devastating outbreak of Russian Aphid.

At the end of the wheat-growing season, Jeremy then contracted me out to spray vegetables in the Gamtoos River Valley. He also arranged for another helicopter to be available for the next wheat season, by basing his calculations on my previous performance. But nobody managed to replicate my achievement again, and he was at a loss to fathom how I had accomplished such incredible results. The overriding reason was that infestations of Russian aphids are rare. Still, I also neglected to divulge that I had calibrated the spray emission for a faster flying speed than experts recommend.

But at the end of the season, I had achieved the impossible. I had logged just under the monthly limit of 100 hours of flying time. I also know that the blessing which boosted my income, was much more than pure chance.

27 CROP SPRAYING IN THE GAMTOOS VALLEY

I was busy spraying against the Russian aphid outbreak in the Western Cape when a couple of men arrived to watch the operation. As I buzzed up and down the field of sad-looking wheat, I noticed one of the men put something on the ground where I was about to spray. I later learnt that they were from a company from the Eastern Cape that contracted farmers to grow vegetables. The person had laid down an upturned pair of sunglasses to gauge the size and spread of the droplets on the crop.

I learnt that they had been pleased with the result when Jeremy Labuschagne called me to come and see him in his office. He informed me that this company wanted Court Helicopters to provide a helicopter and pilot to operate out of a small rural village near Port Elizabeth, called Hankey. He offered the position to me and wanted to relocate our whole family to the Eastern Cape. As an added incentive, he even offered to pay school fees for our children. He also said he would be able to get our eldest son into Greenwood School, which is a prestigious private school, just off the main Garden Route highway, near Van Stadens River Gorge.

By that time as a family, we had begun to carve out a life, and made many friends, in Cape Town. This made me reluctant to move away to what seemed like a predominantly Afrikaans, rural backwater region of South Africa. Although by then, I could get by in Afrikaans, I never really felt accepted, and guessed that I probably tortured their ears when I spoke to them in an intolerable English accent.

I told Jeremy that I would be happy to work out of Hankey if the Company paid for me to live in a hotel there.

I also suggested that each month they pay for a return flight to Cape Town for me to spend a long weekend at home. He quickly accepted these conditions, and I settled into a comfortable routine.

Once there, I mostly had to spray vegetable fields along the rich fertile floodplain of the Gamtoos River, where canals and irrigation pipes directed water to the forty-hectare pockets of land. Each separate piece of land was owned by small-scale farmers, who worked hard as they carried out rapid crop rotation.

These small farms followed the undulating river like a patchwork quilt, and I could not imagine how these simple hard-working folks and their sullen workers could ever get past being subsistence farmers. They also seemed to be at the complete mercy of the contracting vegetable company, 'TableTop', who sold them both the seed and fertiliser on credit, while keeping strict control over the spraying schedule. To achieve this, the company employed chemical reps, who lived in the area, and continually monitored the crops in the fields. They would also organise my daily spraying programme and would choose which chemicals and how much I should use.

After the beautiful large expanse of cotton in Rhodesia and the bigger wheat fields in the Western Cape, it was different spraying these separate small fields. Now, we had to continually move from one piece of land to the next and repeat the whole procedure of drawing a precise, small quantity of water from the nearby riverbank or dam. This water would be pumped straight into the mixing tank, which was pulled around by the orange Court Helicopter truck. After this, we would add a carefully measured quantity of chemicals to cover the selected patch of vegetables. I quickly learnt how to distinguish different crops from the air and developed the technique of delivering the chemicals, in a thick white ribbon of short spray runs, and how to switch off at the exact moment to avoid any wastage.

When I had sprayed the winter wheat in the Western Cape, Edwin who was a brilliant worker from the hangar at the Court Helicopters head office, and Hannes Eksteen, an enthusiastic and hardworking engineer, supported me. Both offered excellent assistance and were such a delight to me during that time. On arriving at Hankey, the local company provided me with a team of five local Xhosa workers to do the work previously done by Edwin and Hannes. I could not communicate with these five recruits in their home language and resorted to my limited knowledge of Afrikaans. None of them had even seen a helicopter up close before. By the look of his bloodshot eyes, one member of this inept team of untrained labourers called Zulu, always seemed under the influence of dagga (marijuana). I also found him frustratingly clumsy and uncoordinated.

On one of the first days in the area, while demonstrating to the local farmers, I became frustrated with these five workers, as they battled to position the mixing tank on its jockey wheel. It was alongside a sharp embankment to draw water from the Gamtoos River below. The comical performance started to amuse the sizeable gathering of the local Afrikaner farmers, who were not forthcoming with any assistance. I became so embarrassed and exasperated by the fiasco of unsynchronised pushing and pulling, that, without thinking, I slapped Zulu with my flat hand across the back of his head. This attention, immediately galvanised the team, spurred them on to a more coordinated effort, and earned wide-eyed respect from the farmers. It also seemed to elevate Zulu to a position of prestige and authority. From then on, he assumed the mantle of leadership and took great pride in personally cleaning the helicopter and pumping grease into the numerous nipples at the end of each day.

Ronald Young, the only English farmer in that area, owned the largest farm near the '*dorp*' (village). His farm was some distance from the river and more expansive than most other farms in the immediate area. He was a good-looking man, who lived with his wife Marge, three children, and his elderly mother, who many years before, had come to the village as the local schoolteacher and married Ronald's father. When I met her, she was already a widow and was blind.

I soon became close friends with the family and started going to Ronald's mother for Afrikaans lessons. This became a time of mutual enjoyment, where she would assign me an essay to write each day when I returned to the hotel, which I would read to her the following day.

Coming from a country where any farm less than one thousand hectares was considered small, for some reason, she took great umbrage after she had asked me to describe the local area, and I wrote, "The farms around here are tiny and only about forty hectares each." She somehow took this as a personal insult.

As she did not have much chance to go for walks, we took the opportunity one day to walk arm in arm around their farm, along the dusty roads and undulating ground, while I described what I saw. In one spot, we came across a cactus plant, they called a 'prickly pear'. I left her for a moment to look at it, when she shouted at me in a panic, as the fruit bristled with clusters of tiny hairs that could cause tremendous irritation and pain, and she thought I was about to pick a ripe one.

After a couple of months of daily visits for my Afrikaans lessons, the hotel manager's wife, who prepared the most sumptuous rich Afrikaans cuisine, casually mentioned that people wondered why I went so regularly to the Youngs' farm. She seemed disappointed to learn that I was visiting the blind mother for Afrikaans lessons as the local community had quickly surmised that I must be having an affair with Ronald's wife. Luckily, Marge was quite amused when I told her about this small-town gossip.

At times, the chemical rep sent me further afield to the Grahamstown area and over the mountain range beyond the town of Craddock. It was interesting to meet farmers in these areas, who were direct descendants of the 1820 Settlers. When some of their ships had run aground near Port Elizabeth, the waves had washed some of these settlers ashore, where it was intriguing how they had become integrated into the Afrikaans community. Several of them had English names, like Roberts or McDonald, but could not speak a word of English, and the only way I could communicate with them was in Afrikaans.

On one of these trips near Craddock, I was spraying a tricky field beneath huge drooping power lines. This took a lot of concentration as I buzzed up and down. Then on the last run, I gently pulled up to clear thick sagging strands, when something subconscious triggered my remembrance about the thin earth wire that stretches across the top of the pylon towers. In icy horror, by pulling maximum power on the collective and pulling back on the cyclic, I just managed to clear the invisible wire, that stretched more than a hundred feet above the ground. This happened on an isolated field in the mountains, and far from any farmhouses or built-up areas, and I shudder to think how long it would have taken someone to come and look for me if had I not returned that evening.

One day they tasked me to fly to the Mossel Bay area. There was a strong gale-force wind blowing from the west, and I realised that the single twenty-gallon tank on the aircraft was not going to be enough for me to make it the whole way. So, I organised for the bakkie to meet me at a pre-arranged spot *en route.* After topping up, the vehicle continued its way, and I took the more direct route to overfly Knysna. However, the wind speed continued to increase, and as I closely paralleled the main Garden Route highway, I was overtaken by cars travelling in the same direction. It got so bad that as the headwind was buffeting the helicopter, I estimated that the actual ground speed was no more than twenty MPH, and at this rate of progress, I would not have enough fuel to reach Mossel Bay. I started to think the bakkie might have passed me, unseen while driving behind the Tsitsikamma Forest trees. So, in the hope of catching them before they reached Knysna, I decided to follow the road more closely, as that would be the furthest, I could hope to get to, before I needed more fuel.

As I arrived overhead the little Garden Route village of Knysna, I realised I would have to land, regardless of whether the bakkie had passed through or not. I would then have to find a telephone to call Mossel Bay and then be stuck near Knysna for a few more hours before the vehicle could return. As the road passed through the little town, I continued to scan the road, and with a leap of joy, I caught sight of the familiar orange bakkie pulling the mixing tank. I was so delighted to escape the anticipated long delay that the ridiculous thought sprung to my mind from the current popular TV series 'The A-Team': "I love it when a plan comes together!"

Everything went well for the next six months, but I started to miss my family, the home cell group, and other church activities with our friends in Cape Town. On one of my breaks back home, I browsed through the Argus newspaper's classified section where I saw an advertisement promising great wealth to their agents who sold wall coating with a lifetime guarantee. The combination of going through 'midlife' crisis and just longing to stay home, prompted me to attend a recruitment talk and to my wife's horror, I was so naïve that I got suckered into the prospect of unbelievable wealth, and I resigned from Court Helicopters.

Later, when I got back to Hankey, Jeremy called me to inform me that Stan Botha was coming to take over from me. He had no crop spraying experience, and when Jeremy asked me how long I needed to teach Stan the ropes, I suggested that he come for two weeks. I thought that should give me enough time to offer him some lectures and enable him to watch the procedure before accompanying me in the cockpit. After then, I would feel confident to leave him on his own. Well, a few weeks later, I received the message that Stan was on his way, and I was to get him up and running in one day. Of course, I felt this was inadequate, but the financial aspect far outweighed any flight safety considerations between Jeremy and Aat Slot, the Managing Director. Despite my protestations, I had no choice in the matter but to comply with their instructions.

I then left flying for a couple of years. But although I felt like a fish out of water, and despite the wall coating agency being a complete waste of time, I eventually did very well as an estate agent. During that time, Stan continued to crop spray for the company, which ultimately expanded to three helicopters that operated in the Western Cape. Later, Stan moved back to fly one of the helicopters spraying potatoes and other vegetables in the Kouebokkeveld, near Ceres' fruit-growing area.

I was sad to learn that Stan had flown into wires while crop spraying and had been killed. Later, the accident investigators discovered that his pockets were full of painkillers, and the autopsy revealed a large brain tumour. This growth must have caused him considerable discomfort and pain, which possibly contributed to the accident.

A couple of years later, another newspaper advertisement led me to contact Jeremy. Thankfully, Court Helicopters took me back, and I started flying again when they initially employed me to return to crop spraying after Stan's accident.

FLYING – BETTER THAN ACTUAL WORK

It was 1984, and after two years of working as an estate agent, without a break or holiday, I was utterly exhausted. Following a few enquiries, Court Helicopters initially welcomed me back to rejoin the crop spraying section I had established just a few years before.

Back to Crop Spraying

Jeremy, the operations manager mentioned that after my first season's phenomenal success, the company had never made the anticipated money. However, that was such a good year because how desperately the farmers needed us to combat the outbreak of Russian Aphids. During that first season, I remained in the cockpit from dusk to dawn and flew six days a week for three months. He also mentioned that none of the other pilots could get anywhere near the number of hectares I had sprayed for each hour flown. A perverse smugness would not permit me to enlighten him that I had stretched things a bit, and instead of calibrating the nozzle's spraying pattern to fly at 60 MPH, I had readjusted it to spray at 70 MPH instead. Apart from this, it was fortunate that I was able to schedule each day's program so that I could move progressively from one farm to the immediate neighbour. This effectively increased productivity by about 17 per cent.

By the time I returned to Court Helicopters, they had also reduced the assistance provided to each spray pilot to one driver. His job was to drive the truck that towed the mixing tank to each farm, mix the chemicals with water drawn from a nearby dam, and then pump it into the aircraft's 'hopper'.

In one of the first spraying jobs after re-joining, my assistant was Piet Muller, a wizened Cape Coloured. I remembered him from when Roger Watt and I had joined the

Company at the end of 1979. At that time, he was employed as a packer in the ship-service hangar at Cape Town docks. His features had that distinctive Bushman look, but I was pleasantly surprised to find him intelligent, and easy to talk to. As we headed off on our trip through Paarl, Wellington, Baines Kloof, and Ceres to the Kouebokkeveld, I quizzed him on how things had been going with the crop spraying operation. He told me that he had worked with several different pilots over the last couple of years, and from our discussion, I managed to get a feel for how things ran now.

On this first job, I replaced William, a New Zealand pilot who had come to work in South Africa and then married an Afrikaans girl. Despite having lived in an Afrikaans community in Stellenbosch for the last few years, he had not learnt to speak Afrikaans. This fact was a stumbling block in an area where many local farmers could hardly understand a word of English. To add insult to injury, he would delegate all communications with the farmers to Pieter. I could imagine this would not have gone well among the conservative farmers in that region, particularly in this isolated mountainous area to the northeast of Cape Town, where they were still a racially biased society. Their only relationship with any coloured folk, was with their workforce called the *"Hotnots"*. This name comes from their ancestors, the Hottentots, an extinct race of people with yellow-skins and bushman-like features. The local farmers had employed their present-day workers, with their families, for many generations now, and the landowners still issued the workers their daily ration of *'Dop'* when they reported for work. However, this kept them semi-comatose but content for the whole day.

While we drove to the first farm, Pieter warned that the farmer I was about to meet was a *"Norse ou bliksem"*; a crusty old sod. So forewarned, I made sure I showed him all due courtesy and respect on our arrival. I introduced myself in my best Afrikaans, which I had acquired over the last few years, and greeted him politely as *"Oom"* (uncle) in deference to his age. Later, Pieter said he could hardly believe this was the same man, who now presented himself as the epitome of a cultured and gracious country gentleman.

After that first week, Petrus replaced Pieter. He was a gentle, well-built Zulu who had previously worked in the Mossel Bay hangar, where Court supported the SASOL's gas rig out to sea.

Petrus and I spent a few months travelling all over the Cape before we operated in the Mossel Bay area for a while. As we travelled long distances together, I enjoyed the relationship that developed between us. It was a revelation to me, as it allowed me a chance to have a glimpse into a black person's way of thinking. Whenever we spread the job over several days or returned to Cape Town for the weekend, I would leave the helicopter at the farm. Our developing relationship was one of mutual respect, and since I was 'the helicopter pilot', I let him drive the 'bakkie' to give him a sense of prestige and importance.

However, what perturbed me, was his limited awareness of the bakkie's speed whenever we drove in fog or at night. I first noticed this when we were heading north of Cape Town, along the main road near Atlanta. A thick fog had rolled in from the sea and reduced visibility to less than thirty metres. Even so, he maintained his speed above 80 kilometres per hour. I held my breath for a few seconds until I became too terrified to restrain myself. So, I instructed him to slow right down to 20 kilometres. Just as we reached that speed, he had to swerve desperately to avoid colliding with a slow-moving tractor, that appeared in the mist.

I thought he might have learnt his lesson from this narrowly avoided catastrophe. But on a later trip at night, he gave me another fright as we returned to the coastal Holiday Inn near the Wilderness between Knysna and George. We had to descend a steep and winding mountain pass to the west of Sedgefield. Once again, I was horrified by his complete lack of appreciation of our speed, as it continued to increase as he battled the steep, downhill, curves that flashed past us. Once more, I had to rein him in.

When we operated in the vicinity of Cape Town, we would return home for the weekend. Petrus would drop me at our house in Edgemead and then take the bakkie to his home in Khayelitsha, the high-density black township, straddling the main road between the airport and the city. Before he left, I would instruct him what time to fetch me early Monday morning. Each time he would arrive about an hour late, so I resorted to setting a time an hour earlier than I wanted to leave. This allowed me extra time in bed, while still hoping that I would not be caught short by his arriving on time. However, one day he was so late I thought he might have had an accident. He then phoned me from a neighbour's house to say that one of the bakkie's tyres was flat and with the extra load on the back, he had struggled to jack up the car in the loose sand of his Cape Flats home.

By the time he arrived, I was upset by the loss of time to get underway to the farm where we had left the helicopter at 'Swartland, in the Northern Cape. While we travelled for a few hours, I was not very communicative. We then turned off onto a very corrugated dirt road, and the rear left rear wheel immediately started to furiously judder. I immediately asked Petrus to pull up, and on inspection could see there was a missing wheel stud. This allowed the remaining four holes to become elliptically enlarged, and without thinking, I seared my fingers by checking one of the studs. Luckily, after recent rains, I was able to thrust my singed fingers into an ice-cold puddle. Since this was the spare wheel, he had only recently changed, we found ourselves stranded in an isolated spot. The only alternative was to creep along the dirt road at a snail's pace. It took over an hour before we could reach the farm, and by the time we arrived, we were already a few hours late. Fortunately, the farmer was sympathetic and understanding, and quickly conjured up a spare wheel and an extra bolt to get us started.

He was a wiry man called "Geel," who had two of the most respectfully polite teenage boys I had ever met. Although he was a slight man, I still remember the vice grip with which he shook my hand, and the love that radiated from his sons' eyes when they spoke to him.

On another trip during the holidays, I took our youngest son Ryan along. He was ten years old then and travelled with me in the helicopter to a farm north of Darling. This belonged to a friendly and hospitable Afrikaans family with three boys and a girl about the same age as Ryan. With insufficient workers to assist in marking the lands, the children took on the responsibility of standing in the fields to help me line up on each spray run. To avoid contaminating them with insecticide, I made sure that I passed alongside them on the downwind side. Now and then, I gave one of the children a chance to ride in the helicopter while I sprayed. They enjoyed the day and the experience of riding in the aircraft as I sped along just above the wheat. When the job was complete, we all piled into the farm bakkie driven by the eldest son, who took us for lunch.

The children happily played football in the backyard of the homestead before we were called in for lunch. This took place in the spacious farm kitchen. around a large yellowwood table, with a roaring wood stove. The daughter, who had cropped hair and was wearing football trunks, looked more like a young boy.

During the meal, the young girl was sweetly attentive towards Ryan, and placed heaped plates of food in front of him. It amused the rest of the family with the special treatment proffered by their young daughter to our sweet-faced blond and blue-eyed English boy. They also commented on how well they had got on, and laughed at how she had sweetly draped her arm over Ryan's shoulder whenever they sat together in the bakkie. Ryan, who could not understand or speak much Afrikaans, innocently asked me why everyone was laughing. So, I told him, "They are laughing at the way Blyda is spoiling you and how she kept putting her arm around your shoulders in the bakkie." With a shocked look of horror on his face and to everyone's amusement, he blurted out, "He's not a girl?!!"

On some trips, Petrus told me stories that highlighted his superstitious beliefs. I was particularly fascinated, when he related how he had once gone to a 'witch doctor' when he had experienced terrible pain in his left shoulder blade. This 'shaman' produced a bisected tennis ball, with which he pushed out the air over the painful spot. After a short while, he sucked out a white crustacean for Petrus to take home in a bottle, and from that moment on, the pain never returned to bother him again.

We had spent a few weekends away from our families while we operated in Mossel Bay. I felt we needed a weekend at home but thought I would pull Petrus' leg, as I said, "I'm sure you don't need to go back to Cape Town this weekend as your wife is expecting her baby quite soon!" to which he responded, "No George, I must go back now to make its arms and legs strong!" To this, I asked, "How are you going to do that? Are you a bicycle pump?" But he misunderstood my feeble joke!

I particularly enjoyed operating around the Malmesbury area. We often stayed at a cosy country hotel, which produced the juiciest and most tender pepper fillet steaks I have ever tasted. I also enjoyed these with French fried chips and a Castle Beer.

The first night I stayed there, I sprayed wheat fields near Darling. Throughout this flight, I suffered a most dreadful and blinding migraine headache. Over the last few years, I had periodically suffered these headaches, mainly if my sleep pattern was disturbed or under stress. Although I think it may have been brought on by a combination of stress and high blood pressure when I had been busy selling houses without a break for months on end. On this occasion, instead of stopping, I continued to spray load after load, while trying to ignore my throbbing head and nausea.

I am not sure how I managed to complete the task, and later, on the drive back to the hotel that evening, I just sat like a zombie while Petrus drove. The feeling of nausea persisted, and my head continued to feel as though it was splitting in two. I could not even talk or have anything to eat at the hotel. The only thing that gave some relief was to kneel for hours in a steaming hot bath with my head submerged under the water while I held my breath. Amazingly, after a fitful sleep, I woke the next morning still feeling weak and shaky but relieved that the splitting headache had gone.

After a few months of crop spraying on the Bell 47 helicopter, the company brought me back to Cape Town docks to renew my currencies on the Bell 206 Jet Ranger and the Sikorsky 58T. During that time, the company gave me many exciting helicopter jobs, which included 'flipping' tourists around the Cape Peninsula.

Moved back to other Helicopters.

Other activities included: flying the lifesavers in the John Rolfe helicopter, erecting masts, photographic and filming sorties, ship service, service to the SASOL rig off Mossel Bay, and lifting massive air-conditioning plants to the top of high-rise buildings. I also took part in a contract for Southwest Africa Broadcasting Corporation. The job was to assist in the erection of a mast on Rossing Mountain on the West Coast near Swakopmund, where I went on two separate occasions. It was satisfying to accomplish this task with the Bell 206 Jet Ranger, with one of the court Helicopter flight engineers accompanying me. We were accommodated in a caravan on the desert sand below the mountain, with an outside ablution block just a short walk away.

I found the soft pastel colours of the desert amazingly beautiful. Because the ablution block was a bit far to walk to when we needed to urinate at night, we installed gutter-type downpipes called 'Desert Lilies' into the sand, where we could urinate into the funnel. However, one night, the alarming onset of an East Wind shook us awake. It came with a roar that wobbled the caravan and sounded like a continuous freight train passing overhead. The wind caused everything not battened down to be either shredded or fly miles away. Despite the intense heat, which made it feel like living in an oven, the windows and door had to be shut tight against the fine penetrating dust, which still managed to get in. One had to brave the gale-force wind to use the desert lilies to urinate, but even though we stood with our backs to the wind, we could not stop what we referred to as "Getting our own back"!

John Clack was the *'rigger,'* who had been sent with his team from Windhoek to erect the large mast on Rossing Mountain. They would bolt sections of the mast together, which we would then take dangling on a length of cable below the helicopter.

John was an unusual character, who most of the other crews avoided. But my engineer, Alan, and I, went out of way to befriend him. We would while away the hours playing childish forms of poker, and although it was only for matches, it was like 'taking candy from a child'. On one occasion we were playing a game which we called Indian. In this game, we would each draw one card, and without looking at it, place it on our forehead. We nominated 'Aces high' and Alan got the Ace of Spades, which was the top card, so I folded immediately. John had the four of clubs, and of course Alan knew that he must have a high card, as I had folded straight away, but John thought he would try and out bluff him. In the end, Alan and I were crying with laughter as John kept raising the stakes.

At that time, I was still fit and tried to do a lot of running when I could. Though it was still hot, when the weather was conducive, I would jog about five kilometres away to the Rossing Mine Club. The engineer would follow me in the bakkie, and we would then have a game of squash, followed by a shower and a swim in the pool. We would end up having a tasty meal at a most reasonable subsidised price. At the club, we made friends with a Yugoslavian called Dragan, his wife, and her two young daughters, who sometimes took us to their home. We also met the most bizarre family where the older lady with a daughter, had married again, and then got divorced. The gentleman then married the daughter, with whom he had a child. The four of them continued to live happily together in the same house, although I am not sure how the child related to each grownup!

Another unexpected bonus was when I heard John Clack and his team speaking in '*Fanigalo*', a Lingua Franca used in the South African mines which we had all learnt as kids growing up in Rhodesia. Unfortunately, with the nationalistic awakening in Rhodesia, this language became forbidden by the underground party. We were, therefore, left not being able to speak to the indigenous in their local language, which we had never had any reason to learn. Now years later, I was delighted to communicate with these workers in a completely different part of Africa.

The nearby town of Swakopmund was very picturesque with its Germanic buildings and history. What was fascinating was the story about how the dolphins would come and frolic among children who swam in the bay, but the moment an adult stepped into the water, they would disappear.

Sent to Mauritius

At the beginning of 1986, Court Helicopters sent me

to Mauritius to crop spray sugar cane from a Bell 206 Jet Ranger. From the moment I arrived, I felt I could live there forever. The most striking features on the island were the sea's indescribable colour, which changed from a light crystal-clear aquamarine inside the coral reef, to an indigo blue beyond. The flowering tropical plants like poinsettia, bougainvillaea and geraniums had rich primary colours.

I had to spray a hormone substance onto the sugar cane which hastened its maturation process. Pernickety, French-speaking, officials were meticulous in organising the crop spraying programme, and always remained on-site to oversee the operation. They would pour the exact quantity of chemical for the field size into the 'hopper,' and insisted that every nook and cranny be covered and yet they were also intolerant of any overspray at the beginning or end of each run. Then, as soon as I had finished spraying the field, they would immediately check to see if any mixture was left over in the hopper. Before this, I had adopted the common practice of filling an extra 5 per cent, with which I could redo awkward tricky corners and where hazardous wires were present.

They were also paranoid about the possibility of rain within 24 hours of spraying, which often caused long delays, as they watched the horizon for the tiniest speck of cloud. When they thought the wind was too strong, it would also cause lengthy delays. They would program me to make an incredibly early start when the wind was at its lowest. But any morning rain would result in their cancelling that day.

One early morning, I thought I heard heavy rain falling outside the window of my little seaside bungalow. I was just about to roll over and go back to sleep, but I decided to peek outside and was surprised that what I thought was rain, was wind rustling through the fronds of a small palm tree just outside my window.

My one delight in Mauritius was the driver, Cyril Dada,

a Creole, who attended to the mixing tank and looked after the helicopter. He had a slight speech impediment which made his English almost unintelligible. What made him even less understandable, was when he sometimes peppered it with mispronounced Afrikaans words and slang, which he had picked up from previous South African pilots and engineers. However, he was such a willing helper, that it was difficult to get too frustrated. I also learnt a lot from him about the local inhabitants of this tiny, overcrowded, tropical paradise, when we were away from the tourist hotels and beaches. One day, he proudly invited me to his small home in a high-density area, more in keeping with two-roomed residences in Soweto. For my visit, his wife and sister went to a lot of trouble to shop and lay on something special, and I was overwhelmed by the extravagantly sumptuous Creole cuisine. I was also humbled when I admired a display shelf in his living room area, which he had proudly adorned with children's plastic farm animals. This adornment was all they could afford as decoration, but he immediately tried to bless me with it as a gift when I showed an interest. It was a moving experience, and I had to decline his generosity with some sensitivity, to assure him of my appreciation without offending him.

I also noted the abundance of heavily laden pawpaw trees in each little garden and was amazed to learn that none of the locals ate them, or the avocados that grew like weeds on the island. I asked Cyril if he could bring me a freshly picked pawpaw every day, which I enjoyed for breakfast.

I stayed in a seaside bungalow, and had a Creole cleaning lady, Jennifer, who also cooked for me. With my smattering of French and Creole, we agreed that I would not tell her what to cook but I would always leave sufficient money on the dresser for her to buy seafood, vegetables, and spices from the nearby market, in the village of Mahebourg. Most of what she purchased was unfamiliar to me, but she, her cousin, and her small daughter, would gleefully hover at the kitchen door for my delighted reaction to each meal.

Each weekend I enjoyed the sound of the locals who partied and sang on the beaches, and I tried to learn a bit about the Island's history. I learnt that Mauritius had changed hands several times over the centuries between the Arabs from the Middle East, and then the Dutch, who had dropped off enslaved people on the uninhabited, tropical island paradise with its abundance of water and plentiful supply of vulnerable Dodos to have as a feast.

Mauritius had gained independence from Britain 18 years before I arrived. It had changed hands several times between the French and English, which resulted in severe maritime skirmishes just off the coast. When the French decided to plant sugar cane there, they imported Indian peasants, as a labour force, and, as had happened in Natal, they also managed to rise above their humble beginnings. When the British were in control, an efficient narrow-gauge railway network ran around the island, with a proper refuse collection. Now all evidence of a railway system had disappeared, and the locals in the villages piled trash on high dumps along the sides of the overcrowded roads. One could only make slow progress on these roads that wound through crowded, corrugated iron, shanty towns, thronged with pedestrians and overloaded bicycles. No one ever bothered to remove the carcasses after vehicles ran over mangy, flea-ridden mutts, which became progressively flattened by every successive set of wheels.

Although English was the official language in Mauritius, most of the locals spoke French or Creole. I had the opportunity to meet some of the gracious French descendants, who had been able to hold onto their magnificent chateaus, surrounded by sugar cane fields and natural forest reserves, which they stocked with deer. The sugar cane barons hunted these animals in a most unsporting fashion. One day a farmer invited me to one of these functions. But I excused myself when I learnt that the procedure entailed waiting on a raised platform, while beaters flushed the beautiful, defenceless animals into a clearing, where they could slaughter them. Although I appreciated the honour of receiving an invitation, I just could not bring myself to go along.

Before I learnt a bit about the local history, I asked Cyril what race he belonged to, and was surprised to learn that he had no idea. The sad thing is, that in this melting pot of humanity, many other races have flooded out the descendants of the enslaved people who had managed to escape. However, they remained happy and straightforward folk, who lived cheaply in the Garden of Eden and least could enjoy the abundance of fruit, vegetables, and seafood. Sadly, they are now lorded over by the Indians and Chinese, who own all the tourist resorts and other trappings.

One day I was asked to go to La Tousse Roc, a plush hotel on the east coast, to take a 'princess,' for a helicopter ride. After completing the day's spraying, I arrived there still wearing my orange flying overalls. As I entered the expansive reception area with its shiny marble floors, feeling conspicuous, I enquired about this princess, who I assumed was from a local beauty contest. The receptionist at the front desk casually replied, "Oh, that must be Princess Stephanie." To this, I jokingly asked, "Is she a real princess?" To which she replied, "Oh yes, sometimes!" Then to my absolute disbelief, I noticed a lithe, elegantly draped figure behind sunglasses appear with her entourage. It was none other than Princess Stephanie of Monaco. She was about to release a song and wanted to be filmed in a video of her running between sugar cane fields, while being sprayed by the helicopter. She was striking but aloof, and after taking her up in the aircraft to the chosen spot, I flew over her while spraying a swathe of water. She had left her beautiful silk scarf draped over the back of the collective lever, but much to my wife and daughter's chagrin, I returned it to one of the people in her group, before flying off.

Over the school holidays, I had stayed on the Island

for nearly three months before my family joined me for a few weeks. The overall beauty, the beautiful beaches, the warm sea, and the friendly people, impressed Tish, and the kids so much, that they were ready to move there permanently. After six years of living in South Africa, the children had not learnt to speak much Afrikaans, but when they thought, they might have to go to school in Mauritius, they learned French words at a phenomenal rate.

Unfortunately, I was far too qualified to while away the rest of my days on a tropical island paradise. So, ignoring all protestations, I was pulled back onto Ship Service operations at the docks, and after Dave Rowe informed the Ops Manager that previously I had been used extensively as a helicopter instructor by the Air Force, I was designated as an instructor on the Bell 206.

So sadly, my days of buzzing up and down as a crop spraying pilot had come to an end.

29 MERCENARY PILOT FOR THE CIA

Casa 212 on Swan Island

Premonition of an extraordinary year ahead

On New Year's Day, I had a premonition that 1986 would be an exciting year. It started with a flourish in February, when Court Helicopters sent me on a contract to Mauritius, where I spent nearly three months spraying sugar cane plantations and taking tourists for flips around the island. Then, I heard that Bill McQuade, whom I had known from Air Force days, was recruiting ex-Rhodesian Air Force pilots, for a contract in Central America. After sending him an application, I felt as though I held my breath for almost six months, while I waited for it to come about.

On the first day of January, Chris Wentworth, an ex-Air Force friend, took over from me in Port Elizabeth, where the company had stationed me on a John Rolfe contract to fly lifesavers over the beaches, packed with holidaymakers. After I had spent two weeks there, the family drove up from Cape Town to join me for the last few days. We then headed off for some leave in Zimbabwe. Before I left, Chris mentioned that Dick Paxton, a mutual Air Force friend, had earned a terrific sum of money flying helicopters to support the Sri Lankan Government to combat the Tamil Tigers. Desperate to make more money, I promptly directed inquiries to the UK's contracting body to apply for a mercenary pilot position. Unfortunately, they rejected my application because I had a South African passport. To overcome this problem, I was enticed to visit the Zimbabwe Trade Commission in Johannesburg to apply for a Zimbabwe passport. I was not optimistic about the application, as I answered N/A to many of the questions that could have disqualified me. To my total surprise, they approved my application, and I collected a brand-new Zimbabwe passport a few months later. This passport proved a real bonus when the Central American contract eventually materialised, especially as other hopeful applicants who held South African passports, became ineligible.

In August, I met up with Bill McQuade in Cape Town, when he made a trip to finalise the recruitment of pilots for the contract. He was still the same laid-back person I remembered when he had joined the Rhodesian Air Force in 1974 to fly helicopters on No 7 Squadron during the bush war. Before that, he had come straight from a tour flying Huey Cobra Gunships in Vietnam. But we later heard that he might have been a CIA plant.

We met at a hotel overlooking Clifton Beach and were well into the discussion before it dawned on me that Bill was lining me up to fly fixed-wing aircraft. This was despite earlier indications, where they had initially wanted to recruit me to fly helicopters. Now, they had earmarked me to drop supplies out of planes into Nicaragua. This change had come about as there were fewer available pilots in South Africa with acceptable passports, and other pilots had withdrawn their applications. Because so many had backed out, Bill asked in his Southern drawl, "Where are all the warriors?" When I mentioned that I had not flown any fixed-wing aircraft for the last nine years, he replied, "Well George, you can fake it!" And with that I was recruited.

Training in the USA

That November, two weeks ahead of the other pilots, the recruitment company sent Thierry Ernst and me to the States to convert onto the Casa 212. We arrived in the USA, on British Airways flights via Nairobi, with a six-hour stopover at Heathrow, where, before catching the connecting flight to Philadelphia via Boston, I took the tube into central London for some sightseeing and reminiscing. On a drizzly night at Philadelphia Airport, we were met by a charming, quietly spoken, grey-haired gentleman called Ed, who drove us to Cape May, in his comfortable brown Lincoln. The traffic moved at a leisurely pace, and as we neared our destination, I caught sight of saloon bars with swing doors, that looked like part of some film sets, and after nearly thirty-six hours, with little sleep, it all seemed as though I was watching a movie. Driving into the small seaside resort, we finally arrived at our little hotel, off four lane traffic, and opposite a long dreary wintery beach.

Once Ed had settled us into the hotel, I was so paranoid about our employers, whom we had already surmised were the CIA, that I could not dispel the feeling that they had bugged both my room and bathroom. I thought they might have installed surveillance cameras, especially as we had been given strict instructions not to discuss the purpose of our visit with anyone before leaving South Africa.

The next day we met the pilot Jim, who would convert us onto the Casa 212. He was a friendly but casual character, who brought some manuals for us, but he seemed a bit of a bush pilot.

The US Navy had established the VLF system to assist its fleet to navigate worldwide. It transmitted Very Low Frequency (VLF) radio waves from nine different stations around the globe and Jim also gave us manuals for the fitted onboard VLF instrument. This was my introduction to computerised equipment, and it took me some time to understand the concept of different pages and how to operate it with its small LCD screen. However, Thierry was already familiar with similar equipment, which proved a great help. His most recent job in South Africa had been as a first officer with Comair, flying out of Jan Smuts Airport, and he was able to explain how to initialise our position and move from one page to the next, enter waypoints, and build up different routes. However, when we started flight training at night, we never had a chance to get hands-on experience.

A week later, Jim arranged to meet us at the Cape May airfield, where we were taken by Ed, after dark. As it was an uncontrolled airfield, we simply took off and flew. As an instructor, Jim was one of the most casual individuals I had ever come across in my flying career, which was immediately apparent by the perfunctory way he pointed out the wings, control surfaces, and undercarriage system by torchlight. On his cursory walk around the aircraft, he said, "Check you have two wings, and the props are still on!" The Air Force and Fleet Air Arm had always emphasised the importance of doing a thorough pre-flight inspection, where we virtually counted every rivet. So, this casual introduction to the Spanish-built Casa 212 aircraft was a surprise.

As Jim instructed on the aircraft at night, I never saw the countryside, except when the moon was out and reflected stretches of, what I imagined, was uninhabited swampland. As we flew, I was also astonished by the density of air traffic. I could see navigation lights all around and above us. But, since they operated on separate radio frequencies, while we flew in uncontrolled airspace, we could not hear radio transmissions between these aircraft and ATC controllers.

After nine years on helicopters, it was strangely familiar to be sitting in a fixed-wing aircraft again. But it still took time for me to adjust to the flatter approach angle when coming into land. While flying helicopters, I often had to land in unprepared places, where wires were the most significant hazards. For this reason, the descent angle, taught by Court Helicopters, was steep. Now, I felt uncomfortable on the Casa's approach angle, which seemed dangerously dragged in!

Thierry and I managed to complete our conversion onto the Casa before the other eight ex-Rhodesian Air Force pilots had arrived. They made up the remaining team members recruited for the contract, which we soon learnt, was to deliver airborne supplies to the 'Contra' guerrillas, combatting the Sandinista Government of Nicaragua.

The operating company were only looking for retired Rhodesian Air Force pilots for two reasons. The first reason was because of their world-renowned reputation. The second was that, still smarting from Vietnam's painful memory, the American Congress had voted against any American citizen operating within fifty miles of the Nicaraguan border.

Meeting with the Boss of Latin Air Services

We soon found out that it was a company called Latin Air Services that employed us, and once the other recruits arrived, the boss Pat Foley, introduced himself to us. He was a short, arrogant, foul-mouthed man, who strutted about in cowboy boots, constantly chewed tobacco, and spat the black discharge into a clear plastic cup. He held the purse strings for the contract, and we later learnt that he had done well by providing air transport to the CIA by Tiger Air, which was his personally owned airline. No one ever acknowledged that we were indirectly employed by the CIA when Ronald Reagan was the President, and Colonel Oliver North had just taken the fall for the Iran Contra Affair.

An instant antipathy developed between the boss and

me. I found him crude and offensive, and my accent and confrontational manner irritated him. He was unaccustomed to being challenged by his staff and certainly not by someone from this motley bunch of newly recruited and expendable mercenary pilots. He was particularly annoyed when I mentioned that under the terms the contract, they were supposed to pay our wages directly into our South African accounts, where our wives desperately needed our first month's salaries at home. Visibly annoyed, he flung a fistful of American dollars across the table in a grandiose flourish, as if to say, "What's all the fuss? It's more money than you have seen for a long time!" This dazzling display was meant to leave us open-mouthed and in awe of his authority and power. So, he did not take it kindly when I pointed out that this money sitting on the table, was no good to our families back home, which made his pudgy cheeks blaze with anger, as he chewed the disgusting wads of tobacco more vigorously. Spitting the masticated dark brown mess into his cup, he snarled, "You sound far too British for my liking."

Failing to leave us confident that he had our best interests at heart, he stormed out in the same whirlwind manner that he had arrived. Once gone, Adrian took over for the rest of the day. He was a good-looking man, in his mid-forties, who, with his blond crew cut and blue eyes, could have passed for a college football player. He quickly won us over with his open, boyish, all-American charm, and the CIA later attached him as their liaison officer at our operating base in the Caribbean.

As time went on, we learnt that Adrian had operated for the CIA as a spy in Rhodesia during the bush war. His knowledge of the *modus operandi* of Rhodesian anti-terrorist operations was alarming and indicated just how much intelligence he had gleaned from loose talk in pubs around the country. Off-duty troops, who gathered to let their hair down, had been more than happy to impress a good-looking fresh-faced American 'tourist'.

As with most of the people we met, we discovered that 'Adrian' was a pseudonym and only at the end of the contract I learnt that his real name was John.

When we met Adrian, he immediately took the opportunity to show off and practise speaking French with Thierry, which he later divulged he had learnt for a two-year diplomatic attachment to Paris.

At this initial meeting, Adrian told us that we would all have to undergo polygraph tests, which were conducted over the next few days by a tall African American, who took us into one of the hotel rooms, where we were connected to his suitcase-sized machine by pipes and wires that transmitted changes to our pulse, blood pressure, breathing rate, and perspiration.

Some months later, Adrian told me that they had designed the tests to weed out anyone who had told other people about this contract, previously dabbled in drugs, or was employed by any other intelligence agency in the world.

During my test, when the operator asked me if I had mentioned this contract to anyone else, I realised that he would ask Dick the same question, which caused me to let out a gasp, and for my heart to skip a beat.

Dick had recently finished the contract in Sri Lanka that I had previously mentioned and told me that he had boasted to some of his non-pilot friends about this new contract. But when I experienced a shock, I knew that this same question would cause Dick to fail the test, and my shock would also have registered on my polygraph test printout. However, although it must have given the testing officer some cause for concern, I was still able to pass the test. Nevertheless, six months later, when I returned from some leave in Cape Town, John diverted me straight to Jamaica to repeat the test. However, it was no surprise when Dick failed the polygraph test with two of our other friends, Eddy and Simon, who were immediately sent home after being comfortably reimbursed.

After all this, and before we left Cape May to operate from the Caribbean, the company advised us that there were now only enough pilots to carry out supply drops into Nicaragua at night. They also told us that the loss of Dick, Eddy, and Simon, left us with insufficient pilots to provide us with helicopter rescue crews.

We had recently seen on the news that the Sandinista forces had shot down a DC-6 aircraft on a similar contract. But I learnt later that the pilots had recklessly dropped supplies into Nicaragua during the day, which had exposed them and made them vulnerable to missiles and ground fire.

I, therefore, give full credit to Vic Culpan, who because of having no helicopter rescue crew available, immediately insisted on a renegotiated package. So, after some lively discussion, our employers acknowledged that it was more productive and cost-effective to get the operation underway and agreed to give us an extra $1,000 danger pay every time we crossed the Nicaraguan border. This extra money would be a bonus on top of our basic salary of $5,500 per month, plus a *per diem* daily allowance. This made it incredibly lucrative during the staggered midterm leave, when the remaining crews had to fly back-to-back sorties each night, and by the time I went on my break, I carried over US$63,000 in cash on me. On my stopover in London, when I booked into my hotel room in Kensington, I was so nervous about being mugged carrying so much, that I hid it on top of the air conditioning vent cover on the bathroom ceiling.

At that time of the year, Cape May, with the rest of the USA, celebrated Thanksgiving, and the shopping malls rang out with, "It's Gonna be a blue Christmas without you", and other Christmas songs. I was fascinated in these shopping malls by the mind-boggling choice of bread and milk in the supermarkets but was quickly disappointed by the sickly-sweet peanut butter and a shiny red apple that I bought to enjoy in my hotel room. The outside of the apple was full of promise until I sank my teeth into its dry cotton-wool pith!

Each day, the temperature dropped rapidly as we made our way around the out-of-season holiday town to frequent the local pubs. It soon became apparent that the local lads viewed this bunch of foreigners, who spoke in strange unfamiliar accents, with undisguised suspicion and hostility.

First deployment to Swan Island

Just as I began to feel settled, and while the other three pilots continued with their flight training, the company summarily dispatched Thierry and me to Swan Island in the Caribbean. This was despite inadequate time and experience being given on the VLF navigation system, or how to deliver cargo during flight.

It took all night to fly the Casa from Cape May to Swan Island, and our last refuelling stop was at Key West, which allowed for the extra distance needed to skirt around Cuba before heading south.

The following day, sunrise was bright as we hummed noisily over a calm sea, and once clear of Cuba in the tropics, we passed overhead puffs of low cumulus cloud. It was difficult to believe that only twelve hours before, we had been enduring a bitterly cold East Coast winter.

We were unable to tune into the beacon on Swan Island, but the VLF navigation system did its job, and we eventually picked up a red return on the radar screen, which indicated the island. When we landed, our eyes stung from lack of sleep, but there was a familiar stirring that warmed my heart as I caught the first glimpse of the tropical island with its white beaches, coconut palms, and casuarina trees. But little could I have guessed that within two days, this aircraft would be sitting at the bottom of the Caribbean Sea!

We parked our plane next to a Dakota DC-3, that was off-loading equipment and supplies. As we climbed out of our camouflaged aircraft, we were met by thick warm air, and greeted by the operations officer. He was an enthusiastic, energetic man in his sixties, who introduced himself as "Tom". Judging by his age and overeager enthusiasm, I surmised that the CIA had brought him out of a boring retirement for this assignment.

Over time, many staff members introduced themselves as "Tom", which we came to realise, were pseudonyms. But I could never understand why the organisation was so unimaginative in allocating names. This 'Tom' had a cocky, strutting manner, like Ossie Penton, who was a well-known Wing Commander of diminutive size, whom many of us had come to know and love in the Air Force. Later, we met Mike, the camp commandant, with another ubiquitous pseudonym. He was tall, with messy fair hair and something quite seedy about him, although I could not put my finger on what it was, except that he seemed to lack any pride in his appearance.

They gave us a few days to settle down on this small isolated tropical island paradise situated south of Cuba. The US government had formerly used it to beam radio propaganda to Cuba's people and as the launching pad for the failed invasion into the Bay of Pigs in 1961, after Fidel Castro had assumed power.

Now, it was a Honduran naval base, which a letter written by Christopher Columbus, proved that it belonged to them. However, the little harbour and jetty could only take small boats.

There are two islands. The main one is about two miles long, with the length taken up almost entirely by a grass runway. The other island, lying off to the east is just a volcanic outcrop of sharp jagged rocks inhabited by sea birds.

We found a cluster of flat-roofed buildings on the

western side of the main island, which were remnants of better days during the US occupation and now used by the Honduran navy. There were also two other houses near the harbour. Mike, the camp commander, lived in one, and Cyril, who came from the Cayman Islands but spoke in a broad Jamaican accent, lived in the other. He rented the house from the Honduran government and used the island to graze a herd of shiny brown cattle. I immediately warmed to him and enjoyed his brogue, which was not unlike some broad rural English accents I had heard. But after talking with him among a group of American storemen, one of them asked me if he was speaking Spanish and was surprised when I told him that Cyril was speaking English.

The ops manager took us to an open billet in the Honduran Navy section, where I put my little black trunk at the base of one of the beds on which, wearing only a pair of shorts, I slept happily for a few hours. Remembering the recent icy weather in Cape May, it was liberating to walk around the camp wearing slops, and even have a swim off a picturesque beach. Although there was no coral reef, the underwater rocks, afforded protection against more massive sharks that approached the bathing area. Later that day, we did some snorkelling with borrowed masks and spear guns.

Ditching in the Caribbean

Although Tom was the operations officer, they had not yet established an ops room on the base, which was still being set up. Being impatient with the pace of things, and eager to commence supply drops to the US-sponsored Contra guerrilla force, Tom persuaded us to carry out a night reconnaissance flight to the Nicaraguan east coast.

His decision was an impulsive one, and he did not bother to liaise with the dishevelled 'camp commander' who showed little interest in what we did anyway. Despite the whole operation hinging around our activities, Mike's heart did not seem to be on the job.

That afternoon, we practised a few circuits to re-familiarise ourselves with the aircraft and then planned the sortie. To further compound the impending disaster, at the last minute, Tom decided to join us without making any arrangement for someone to man the operations radio while we were gone.

When we took off, we headed off into a clear night sky and on the way out, the VLF responded well. When we reached our calculated and programmed waypoint, we turned around to return home by following the heading and distance shown on the VLF. At about 200 nautical miles out, a picture showed up on the radar exactly where we expected the island to be. However, we still had no response from the NDB to confirm that we were heading towards Swan Island, and nobody was set up to respond to our radio calls on the operations frequency. However, as we set up our descent at the pre-planned distance, this gave us no reason for concern.

The first clue that all was not well was when we levelled off at 2,000 feet above the water. The VLF indicated fifteen nautical miles out, but I was surprised that we could not see the 'gooseneck' flares lighting up the sides of the runway. Suddenly, heavy pelting rain assailed the aircraft as we entered an unexpected thunderstorm. Our hearts sank with a sense of foreboding as it dawned on us that what we thought was a picture of the island that showed up on our weather radar in the 'map mode', was a thunderstorm cell. Trying to keep calm, we started a square search pattern and went in and out of one storm after the other, with no way of knowing if one of the storms was shrouding the runway. Still unable to contact anyone on the radio, the situation quickly deteriorated, with the aircraft at the low altitude rapidly using up fuel, at an increased rate of consumption. It became apparent that our options were diminishing quickly, as it was like looking for a needle in a haystack and we started to pray that we might still see the runway before the fuel ran out.

While we searched for the island, I concentrated on the flying to keep the aircraft from descending below 2,000 feet. At the same time, Thierry scoured the aerial chart to find any operating VOR/DME beacons in Central America or Cuba, to give us a 'fix'. The VLF, which was our only available guide, had given us a false indication of our position, and the island's NDB, which should have helped us to home overhead, was not functioning. NDBs can also prove unreliable in the vicinity of thunderstorm activity, but if anyone had responded to our radio calls, it might have helped if they could have fired a flare into the night sky.

As we continued our search pattern, I was reminded of previous accidents resulting from a lack of attention to the flying instruments, especially one incident that happened when the crew of an Eastern Airlines Lockheed L1011 Tri-Star was flying over the Everglade swamps. All the crew members had focused their attention on a warning light that showed a problem with the landing gear, and as the crew put the aircraft into an orbit on autopilot, they failed to notice that the aircraft was slowly losing height and they flew straight into the swamp. There were 101 fatalities and 75 survivors.

I also kept an eye on the fuel gauge as it continued to drop. What concerned me was that we might run out of fuel and would have to ditch into the sea from a glide approach. I knew that if we stood any chance of survival, we would need to ditch the aircraft while we still had engine power. I mentioned this to Thierry, as although we were both checked out as captains, he was the captain for this flight.

When it became apparent that we would not find the island, and as the fuel tanks were almost empty, we decided to ditch the aircraft while we still had power. I suggested we return to an area where we had seen a vessel heading southwest. When we located it again, we flew low over it while

I flashed the landing lights in Morse code for SOS, although we learnt later that they had not recognised the signal. Switching on our landing lights and all other navigation lights, Thierry positioned the aircraft to land across the ship's bow, in the hope they would see us. Meanwhile, in preparation for the landing, Tom left the jump seat and went to brace himself against the bulkhead wall behind the cockpit.

After tightening our seat belts, Thierry ditched the aircraft into the Caribbean Sea. Thankfully, although there was no recommended procedure, there was not much swell, which led Thierry to use the same speed and approach technique he would have used when landing on a short airstrip. He also knew how difficult it would be to successfully ditch an aircraft with a fixed landing gear so he held off just above the water for as long as he could.

When the aircraft finally hit the water, it came to a lurching stop, and seawater gushed in from the front of the cockpit floor and swept my kitbag to the back of the cockpit. However, by stretching back, I managed to retrieve my bag, which contained my emergency equipment and passport. I then opened the hatch on the top bulkhead behind the cockpit, and the three of us clambered through it onto the wings and, while hanging onto the HF aerial, we inflated our life jackets. Fortunately, the aircraft was kept afloat just above the water level by the buoyant empty fuel tanks inside the wings. I pulled out a pencil flare from my bag and shot it off into the night sky. Thankfully, the ship crew had seen the aircraft splash into the sea and were circling around us. I took out an emergency light and broke the tube to start it glowing, and luckily, they could use this single strip of luminous green light to see our position clearly enough for them to launch a rescue boat.

Rescued at Sea

The rescue boat managed to come up alongside us, and

without even getting our feet wet, we transferred from the aircraft wing to the small boat. Later the joke was that when I stood with my bag in one hand and held onto the HF aerial with the other, I looked as though I had been waiting for a bus.

The ship turned out to be a Norwegian vessel, and the ship's captain courteously welcomed us aboard his cargo ship. Thankfully, I had stored some telephone numbers on my Casio watch, and the radio officer was able to contact Ed, the elderly gentleman who had looked after us in Cape May. I explained our predicament to him and told him where the ship was heading and its ETA. The captain then arranged for us to spend the night in the ship's sick bay.

As we entered Puerto Cortes the next day, a dark green Bell 202 helicopter landed just outside the dock area to collect us. The radio officer, who was the only person on the ship who had shown any suspicion of us, refused to return our passports, saying they would have to be handed to the immigration authorities when they returned from lunch. To avoid further unnecessary complications, I sought out the ship's captain, who at once ordered the release of our passports, and with them back in our hands, we just walked off the vessel.

The ship had docked just before the lunchtime siesta, and it felt like a James Bond movie as we casually walked through the unguarded security gate to board the waiting helicopter. The pilot then took us south to Tegucigalpa, where we stayed the night in one of the US Embassy's safe houses. The city lay in a bowl of surrounding hills, and as we overlooked it that evening, we had a fantastic vantage point as we watched a spectacular firework display. This was interspersed by bursts of tracer from automatic weapons as the population exhibited how they celebrated the approaching Christmas festivities.

The following day, we were taken on a scenic drive and had a chance to climb up the side of one of the Mayan pyramids.

Then on our third day, we were delivered back to Swan Island in a twin-turboprop Dornier aircraft flown by Dan, an ex-Marine Hawker Harrier pilot. While we chatted with him, he told us in his amusing way, how his career had abruptly ended after he crashed a Harrier 'jump jet' at an air display, right in front of a crowd of spectators. He described that while he was in hover flight near the ground, he sharply dropped the aircraft's nose, as if bowing to the spectators, but his overall sleeve caught the thrust vector lever. As the stricken plane crashed, the ejector seat shot out and hit the ground with him still strapped into it. Later, as he lay in hospital with multiple injuries and broken bones, he was interviewed by a board of enquiry members. During this investigation, it did not go down well when they asked him what had caused the accident, and he replied, "My pappy always said, 'If you can't be good, be spectacular.'" He was a short stocky man with a slow drawl and was a flamboyant exhibitionist. In time we referred to him as, "Dan, Dan, the action man", as he always entertained us when he paid a visit to Swan Island. Each time he approached to land, he would skim the treetops at the runway threshold and bring the Dornier to a spectacular stop, in less than 20 metres.

Return for more training.

After we were delivered back safely to Swan Island, it felt strange to walk into the billet and see the trunk with my name on it. It looked so familiar as it sat innocently at the bottom of the bed, with my slops on the one side, just as I had left them! It was an eerie thought that if ditching the plane had killed, this is how they would have been, waiting like this, for my return.

The day after we got back to Swan Island, a board of enquiry team arrived with a US Air Force liaison officer, also called "Jim". Later that evening, when the interviews were over, I had an opportunity to talk to him. But it took me somewhat aback when he openly informed me that he did not want to know any of the pilots on this contract, as he felt our life expectancy was limited. However, at the end of the contract, I had another opportunity to meet up with him during the last week before leaving the island to return to South Africa. On this occasion, there was a complete change in his attitude, as we sat in a secluded cove on a small beach, soaking up the softening rays of a glorious tropical sunset. After we had snorkelled, he was now more than happy to inform me that his superiors had been "dazzled" by our successes and the professional way we had undertaken the operation over the last year. He said what had impressed everyone was our pre-take-off briefings and debriefs, especially as everyone could set their watches by the precise adherence to our briefed start-up and take-off times.

After they finalised the board of inquiry, Thierry and I were flown back to the USA the day before New Year's Eve and taken to join up with the other three pilots at a motel next to the Wilmington Airport, where they had just enjoyed some time after they had completed their conversions. The next day, we celebrated New Year's Eve at one airport bar. After I arrived there, a young lady asked me what the time was. So, I said, "Half past eight." To which she replied, "I like that." I then asked, "What do you mean?" She answered, "I like your accent, we say eight-thirty." To which I responded, "I like your accent." To which she replied, "*I ain't got one.*" It was an appealing pub with World War II memorabilia hanging on the walls and ceilings. The toilette area had sandbags against the walls with loudspeakers blaring out Winston Churchill's wartime speeches.

But with the memory of our ditching into the Caribbean, still fresh in my mind, I consumed far too much alcohol at this celebration, and the next morning, at breakfast the other pilots tucked into greasy eggs and bacon, while I nursed a blinding hangover, which was a source of amusement to them all.

The company then moved us to Norfolk, Virginia. This transfer was done under the care of a modest man, also called Jim, who had lost one leg in some incident during the Vietnam War. We stayed in a motel on the city's outskirts for a few days before relocating to Charlotte, North Carolina, where we studied for two days before writing the American ATPL examination, which we needed to fly American-registered aircraft. While in the high-rise hotel room in downtown Charlotte, I studied for 25 hours over the next two days by going over the book listing 1,200 multiple-choice exam questions. Once we had all written and passed with flying colours, we flew, in the remaining two Casa aircraft allotted to the operation, to Van North Airfield near El Paso, Texas, close to the Mexican border. This allowed us to get some hands-on practice with the VLF navigation equipment during night navigation exercises, where we learnt how to carry out night supply drops from 800 feet AGL. It was in the semi-desert countryside, reminiscent of the Karoo sheep farming district around Sutherland, northeastern Cape Province, full of prickly cacti and mounds made by prairie dogs that looked like the South African meerkats.

In the evenings, they treated us to thick cuts of delicious, tender rump steaks cooked in a sawdust smoker. While here, I celebrated my 45th birthday by tearing around on a quadbike with Mark Fenner on the pillion seat. He hung over my neck in hoots of laughter, every time I changed gear as I unwittingly applied full brake on what I thought was the clutch lever.

Vic Culpan had recently flown the C130 aircraft in the SAAF and explained the procedure they had used in South Africa. While here, we perfected the technique of carrying out low-level supply drops, by incorporating the same method, where we made one low-level pass over the drop zone at 800 feet AGL, followed by a teardrop procedure to bring us back in the opposite direction. During this manoeuvre, we would open the cargo ramp door during the turn and then dispatch the load just before we got overhead the DZ (Drop Zone). The pilots made the call over the intercom to the 'kicker' (dispatcher) and, if it were well-judged, the load would parachute down to the target area on its pallet.

On my navigation exercise with zigzag dog legs at several pre-selected waypoints, everything went without a hitch. Then, after making a successful drop, we set course to return for our final landing at the airfield. But when we arrived at the indicated waypoint, there was no sign of the runway lights or the lights of Van North. After the initial shock, we surveyed the area and realised that there had been a total power failure in the area. We eventually managed to locate the runway by the faint light of the moon, and the only option left to us was to carry out an utterly blind circuit on instruments and hope that we would line up on the final approach to the runway.

Once on the descent towards the threshold, I used the landing lights to make necessary landing adjustments. Unfortunately, not having flown fixed-wing aircraft for the last nine years, I was unaware that my final approach was more like a dive bomb attack onto the threshold, which required a hard flare to level off just before touchdown. As this was the last landing for the night, all the other pilots were on board. Initially, they stood at the open cockpit door, as the technicians had removed all the seats to configure the aircraft for supply drops. But, when they saw the terrifyingly steep angle of approach, anticipating a crash landing, they scuttled away to brace themselves against the cockpit bulkhead. Fortunately, I managed to avert a catastrophe at the very last moment, by hauling back on the control wheel.

As we taxied off the runway, they were back at the cockpit door, amazed that I had pulled off the impossible, but it was too much for them when I innocently turned to Mark, in the co-pilot's seat, and understated, "I think the approach was a bit steep!" This caused the onlookers to burst into peals of raucous laughter, which I am sure was from the after-effect of their recent surge of adrenalin.

After this training period, they sent us to Orlando for a week, where they treated us to Cape Kennedy Space Centre, Epcot Centre, Disneyland, and the Water World. I remember the drive out to Cape Kennedy as a complete culture shock. As we drove away from the outskirts of Orlando for over an hour, we passed a continual string of fast-food restaurants and other tourist traps that lined the road and bombarded my senses. It was so mind-boggling that in the end, my brain could not take any more of it. It felt like a calculator that had not been designed to cope with the size of the calculation and displayed 'e' for 'enough!'

Return to Operate from Swan Island

The next time we flew was to ferry the two remaining aircraft allocated to Swan Island with Jim, the likeable instructor, who had assisted with our recent training. He was extremely encouraging and, despite having lost one leg, like Douglas Bader, he did not let that deter him in any way.

One day he asked me if I knew Cocky Beneke, who was now flying helicopters for the US in Central America. I said I knew him well and that he was legendary amongst his Air Force compatriots and was also awarded the Silver Cross from the Rhodesian bush war. I also commented that Cocky was a "real character", to which Jim gently replied, "I think you are **all** characters!"

Swan Island has a coastline of sharp black larvae rocks and little coves of white sandy beaches. The northern side has a long straight beach, lined with coconut trees, which I believe had been planted as a commercial enterprise but were largely wiped out by a recent tropical cyclone. There were also areas of thick forest and open grassy areas, where the cattle had done such a good job that they had made them look like manicured lawns.

When we returned to Swan Island, we noticed a half-finished structure near the aircraft parking area to accommodate us. We immediately stripped off our shirts and got stuck in with hammers and nails.

In no time, we finished putting together the wooden frames, plywood partitioned bedrooms, kitchen, crew room, showers, and toilets. There were no windows, but that did not bother us as we would be sleeping during the day and flying at the night. On the plus side, we enjoyed the luxury of air-conditioned units. However, we were not given a hot water geyser, as Adrian, who had replaced both Mike and Tom, the former camp commander and operations manager, had a bee in his bonnet about our exorbitant earnings. He felt we should at least rough it to compensate for earning much more than he or his staff did!

Behind our accommodation, the builders put together open billets for the Nicaraguan 'Contra' soldiers, who the Contra leaders had sent from the field of operations for a period of R&R. They also had to assist Johnny to look after the two Casas, offload cargo brought in by the C47 planes and help pack the palettes, which we dropped from our aircraft. There were also storemen, operations assistants, cooks, cleaners, and kickers who dispatched the loads from our planes to the Contras.

A tunnel building, covered with a green-painted corrugated iron roof with gauze-netted doors, served as the dining hall. But Thierry was disgusted with the standard of food, which we initially shared with the troops. Their food consisted mainly of cooked dried beans and rice, which the workers relished with a lavish sprinkling of Tabasco sauce. They would then gobble it all up with noisy zest while leaving excessive scatterings of rice on the floor.

After endless complaints from Thierry, who is an excellent chef, we were eventually allowed to put in grocery orders, including blocks of cheese, good cuts of meat, and fresh vegetables. Due to my lack of culinary expertise, I was only trusted to peel potatoes and cut the vegetables, which victuals Thierry would then conjure up into more palatable and nourishing meals.

We had an American aircraft mechanic called Johnny, who was a rough diamond, who cared for the two Casas. He always carried a small pistol in a holster inside his cowboy boots, wore a baseball cap and worked on the aircraft without a shirt. It was amusing how he handled the Contra workers allocated to assist him with refuelling and maintenance tasks, especially as he could not speak Spanish. When I asked him how they managed to understand him, he said, "If they don't understand, I just holler!"

Although he looked pretty rough and dangerous, I suspect he was soft as jelly inside. One evening, Mark Fenner, who was one of the other pilots, and always said I did not know my own strength, challenged Johnny to an arm-wrestling match against me. Well, I reluctantly obliged, expecting to be crushed, to my surprise, he was not as strong as I had imagined, and I was able to beat him with both arms. When we started, I remember his eyes opening wide with shock as it dawned on him that he could not budge my arm from the start position. But, although I know some geometric tricks, I never had to use them. After that, when I went off to bed, Johnny, and Mark, sat up talking, and while Johnny had a few drinks, all I heard for the next few hours from him was, "That's one strong ...*adding another commonly used expletive*".

We soon got into the habit of staying awake until midnight, watching taped films from a VCR library and playing Trivial Pursuit. Apart from occupying himself with the menus, Thierry spent long periods on his bed, reading the Air Almanac, which gives sunrise and sunset times and the moon phases. I was never certain whether he lacked more stimulating reading material or just wanted to ensure he never flew on moonlit nights.

We often went spearfishing among the sharp volcanic rocks and after a while became quite blasé while swimming among menacing five-foot sharks or being escorted by a school of sharp-teethed barracuda. One day, Vic called from the other side of the rocks to say a tunnel led to deep water on the other side. However, after we found, and swam through it, we could not see Vic anywhere. Then we heard him frantically calling us on the land side of the rocks, and wondered how we could have missed him as we passed through the narrow tunnel, until he told us that just as we were coming through to join him, he came face to face with a giant shark. To escape in a panic, he ran straight over the razor-sharp rocks while still wearing his flippers.

When we had no night flights, we would sometimes search in the afternoons for enormous crayfish, which the Americans call lobsters, and other edible tropical fish, to supplement our provisions. This seafood was a real treat when cooked fresh from the sea.

During our stay there, I got into the daily habit of running a circuit along footpaths around the island. These led me past long sandy beaches lined with coconut palms and casuarinas, then through patches of tropical forest and secluded coves. After months of this daily routine, I became exceedingly fit and ran the six-kilometre course at an ever-increasing speed. But at one stage in the year, it became difficult to avoid the hundreds of pink crabs, with protruding eyes and raised front claws, as they ran sideways, to and fro, across my path. In the end, my only choice was to ignore them and just feel them crunching beneath my running shoes.

Despite flying until just before dawn, most of the other pilots still woke up at about 1000 hours while I continued to sleep. However, they were very considerate towards anyone who needed to sleep and did not speak above a whisper. Despite this, I slept with eye patches and earplugs, which I had kept from previous commercial flights. These aids helped me sleep until 1600 hours. When I woke up, I would go to the briefing in the operations room at 1700 hours to check if I was on the programme to fly for that night's drop. I would then go for my run, followed by a swim, or spend some time exploring the forest on the island with Mark.

During these bushwalks, we were often startled out of our wits by green iguanas, that would wait until the very last moment before bursting from the undergrowth, in an explosive run on their powerful back legs. The saddest thing was to witness the Nicaraguan and Hondurans chasing down these defenceless creatures, which they told us they relished as a delicacy, proclaiming that they tasted just like chicken.

Although at that time, Mark was twenty-eight, and I was now forty-five, we struck up a remarkable friendship and enjoyed exploring the island together. We would attach wire hoops to the top of long poles to pull down the abundant young coconuts, which we would chop open with our issued aircrew knives to get to the delicious, fragrant coconut milk, before relishing the succulent and floppy flesh. At other times, while the other pilots languished around the crew accommodation, we would make up paper chases for each other around the island.

I even tried to grow vegetables by using empty ammunition boxes and fertilising the soil with cattle manure and my activity would often attract the curious attention of some Nicaraguan onlookers. One day, I asked one of them if he would mind collecting more cow dung for me, and when he asked, "*Cuando?*" He was relieved when I said, "*Mañana!*" As in their parlance, this was virtually procrastination, which meant it would never happen. However, in the end, the iguanas got to the plants before I ever had a chance to reap any of the crops.

One night, Mark and I flew together on a sortie with the so-called 'Special Forces', where their seniors had tasked them to parachute into the operational area. When we reached the drop zone, there was some problem in the rear of the aircraft, but as the kickers only spoke Spanish, we were unsure what the excitement was all about. I suggested to Mark that he should go and see what was happening. When Mark returned to the cockpit, I learnt that one of the paratroopers had refused to jump out. The terrified soldier just stood at the opening with a vice-like grip holding onto the side of the aircraft. When I asked Mark how he had solved the problem, he casually informed me that he had just slapped him, and then pushed him out with a boot to his backside. I was a bit horrified when I heard this and expected some repercussions but later when I spoke to one of the kickers, in awe he said, "That Mark! He is very brave!"

One day, one of the more senior Nicaraguan personnel on the island, who was either recuperating from injury or on R&R, asked me what our mother tongue was. When I said English, he could hardly believe me. During briefing and debriefing sessions, I often noticed the blank glassy-eyed expressions whenever we broke into colloquial slang from the Rhodesian bush war. In disbelief, he asked, "What about Vic?" and when I assured him, that he was also English speaking, he said, "But he doesn't say 'boots,' he says, '*beets*.'" I then realised just how different we must sound to their ears, which were more accustomed to the American accent. One day, after that conversation, a lady came to present an intelligence briefing, and during it, Vic asked, "What do you do with your wounded?" This question took the briefing officer aback as she queried, "Our *winded*?" On another occasion, while playing Trivial Pursuit with Carl, who had been a former US representative in the Tokyo Olympics karate team, the question put to him was, "How many colour cards are there in a deck?" To which he asked in astonishment, "In a *dick?*"

Vic ran our roster, and after receiving the position for that night's drop, the duty crew would set about planning a zigzag running route to the DZ. For this, they would select prominent mountain peaks as waypoints and calculated the total distance and time to arrive at the planned drop time. To provide sufficient safety clearance above the terrain, we would add 1,500 feet above each mountain peak.

Then during the flight, we would use the weather radar, on which we selected the pencil beam setting, and tilt it right up to only reflect the mountain peaks, which we would use as a cross-reference against the VLF navigation system. We would also tune into the VOR/DME beacons, which the Nicaraguans (our enemy, so to speak) kindly and accommodatingly kept operational. We would then switch all our external navigation lights off and select the cockpit lights to dim red, and never flew during the full moon. As planned, we would keep 1,500 feet above any point along the route, except when we ran into the drop zone, which required us to descend to 800 feet above the DZ.

When we renegotiated our contract, we made it clear that we would always arrive overhead the DZ at the exact pre-arranged time, and after the troops waiting for their cargo had transmitted the expected 'authentication word'. By following this procedure, we could ensure that their position had not been overrun. Their position would be further verified by their laying out the fires or flares in a pre-arranged pattern. We would then make a single pass overhead and carry out a thirty-degree turn away from the run-in heading, and after one minute, we would return to overhead on the reciprocal heading, for the actual drop. But, if the DZ's ground forces were not ready when we flew overhead their position, we would just leave the area and return to base.

After a short time, a young Nicaraguan whose *nom*

de guerre was 'Napoleon' joined the team as our onboard interpreter. He was eighteen, with a stocky build, classical good looks, and shoulder-length black hair. During the run-ins to the DZ, our interpreter would call the troops on the ground to confirm that they had everything set up and that the team was ready for us to carry out the drop. Once we saw the DZ lights, he would transmit a pre-arranged word and wait for the correct reply before we came any closer. On one occasion, Napoleon spent some time chatting on the radio before informing us that it was not safe to continue with the drop, so we returned to base. At the debrief, we learnt that the person answering the radio had not responded with the correct authentication. He had pretended to be the local area commander, under whose command Napoleon had served and knew well. To be entirely sure the man on the ground was not who he said was, Napoleon asked him what his girlfriend's name was, which of course he did not know. Then later at the debrief, Napoleon's account of the story was a source of amusement to the Spanish-speaking members in the operations room. Later, when I asked what was so funny, I was told that when Napoleon informed the person on the ground that he had not given the correct reply, he just said to Napoleon with some frustration, "Just come down, we are waiting for your drop!" To which Napoleon replied, "The only thing coming down are your mother's drawers!"

Sometime later, Napoleon came to me for some advice that related to a shocking story about him and a friend on a recent spell of leave. He said they had run into a bit of bother with a police officer in Tegucigalpa, that culminated in them killing him in a back alley. Napoleon's main concern was that his friend could finger him as the culprit. However, when he returned from his next leave, I was stunned to learn that he had ignored my advice to be cautious and not do anything hasty. Without any show of remorse, he happily mentioned that he had sorted out the problem by hiring a 'hitman' to kill his friend and produce the gold fillings from his teeth as proof. After that, he went on with total peace of mind!

In Rhodesia, I had grown accustomed to enemy fire while flying Alouette helicopters during anti-terrorist operations. But nothing had prepared me for the long streams of tracer fire coming straight towards the aircraft on many of those night sorties. It was like being sprayed by a jet of water from a garden hose. Fortunately, the probing streams of tracer veered off behind us at the very last moment, but although we sat on beefed-up armour-plated seats and wore flack vests, which gave us some sense of security, we still took evasive action by turning away from the firing position. Auspiciously, the enemy directed their fire towards the aircraft's sound, and although the plane was not so fast, it was still slightly ahead of the engine noise. On some nights, it was as though we were running the gauntlet when the Sandinista occasionally aimed heavy weapons at us and the danger of being shot down was ever-present. I later calculated that on average we came under heavy anti-aircraft fire once in every seven sorties. However, the realisation that the danger-pay for each trip exceeded my gross monthly salary with Court Helicopters, made it seem worth the risk.

Our primary source of concern was the chance of encountering heat-seeking anti-aircraft missiles like the Sam 7 (handheld Strella missile) and the Sam 2 (large vehicle transported Missile). We knew the Nicaraguan armed forces possessed them but hoped that intelligence sources had tracked their location properly. But our not being US citizens left the nagging feeling at the back of my mind that in the final analysis, we were considered dispensable, and I was sure that the Americans would flatly deny all knowledge of our involvement in supplying the Contras.

On some nights, we would see a wall of thunderstorms, which flashed in continuous and erratic staccato across our track, and, not having the luxury of autopilots, the aircraft always had to be hand flown by reference to the flying instruments. So, to avoid excessive fatigue, we took turns at the controls and steeled ourselves for the bumpy ride, while we penetrated a line of cumulonimbus towering up to 60,000 feet. As the Casa did not have pressurisation or oxygen, to avoid anoxia, which has an insidious way of sneaking up unnoticed, we could only fly safely up to 10,000 feet. On one drop during a heavy rainstorm, even though we could not see any indication of the reception party's position but had dropped at the same place the previous night, we felt confident enough to descend to 800 feet AGL in the valley. After receiving the correct authentication response, we flew overhead the DZ, and we asked the ground forces to call us when they heard the aircraft directly above them. We then called for the kickers to dispatch the palettes when we received the call. As we said goodbye to the ground forces, they told us that one of the palettes, that came hurtling out of the heavy rain, was an almost direct hit and narrowly missed one of their troops.

Returning for some leave

After being on the island for a few months, we started taking our allocated month's leave, which meant that the remaining pilots had to fly more often. Some nights we had to make two trips, and in one month, I flew thirty-one sorties, which was quite some going, considering that most trips were nearly six hours long. Each time someone went on leave, we had a chance to fly with them from the island to the Cayman Islands, to enjoy a night away in Dan's Dornier aircraft.

The first time we went to the Caymans was quite a treat, with a chance to relax and act like tourists. We were fascinated to see the quaint wooden homes with brightly painted low picket fences and pretty gardens. It was also marvellous to relax on the long Seven Mile Beach and have a chance to do some windsurfing without the continual expectation of having to do a drop that night and the unspoken dread of not making it back next time.

The first night, we stayed at a comfortable small hotel on the beachfront, and after having something to eat, we all sat around the pool to unwind with a few drinks. But in no time, I found hordes of mosquitoes devouring me. None of the others seemed to be affected by them, and I wondered if the alcohol consumed by them, acted as a repellent. But it soon drove me to distraction as I kept slapping at the mosquitoes with bumps coming up on every exposed bit of flesh. All I could think of was the round tin of Zambuk (a South African remedy) in my room, which I always carried with me and knew the green eucalyptus-smelling paste would soothe the itches. So, I quickly excused myself and escaped to the luxury of my air-conditioned room to watch a bit of television while the Zambuk did its trick. This same ointment healed all manner of scratches and grazes. Initially, it had amused the others whenever I produced it. But after a while, they would repeatedly ask me for some, especially after we found it was the only thing that healed grazes caused by brushing against the razor-sharp, underwater rocks when spearfishing.

When my turn came to go on leave, Tish was surprised to see me sporting a beard and looking quite pale, despite having come from a few months in the tropics. It was a wonderful feeling to return with some excess money in my pocket to splash out a bit. I immediately arranged for contractors to build a swimming pool, a garden wall, and open one bedroom with a French door onto a new patio. Tish also had fun buying shrubs to fill our little garden in Edgemead.

The one thing I avoided was contacting any of my flying friends, as I knew it would be hard to avoid telling them about the contract, which was supposed to be top secret. It was fortunate that I had kept a low profile, as the moment I landed back at the Caymans, Adrian met me, and although I could see the other pilots waving at me, I had no chance to speak to them, as they were immediately whisked back to Swan Island without me. Separating us was a bit confusing at the time, but they later told me that Adrian kept them away from me to prevent them from warning me what was about to happen.

I was put up in a comfortable hotel on Cayman Island, with a well-stocked bar fridge and relished the selection of exotic cheeses, chocolates, and drinks. The next day Adrian and I boarded a local airline, which took us in a small aircraft to Kingston, Jamaica. When we got there, Adrian went straight through the clearance for diplomatic personnel, while I felt stifled amongst a horde of passengers in a hangar-like clearance hall. In the end, it took ages to clear me through customs and immigration.

Once again, Adrian booked me into a different but very friendly, three-star tourist hotel, while driving himself to a more luxurious one. Once settled in, he contacted me to say that my previous polygraph test had been inconclusive, and the same big African American testing officer would conduct another one. I was not unduly concerned and spent time listening on my new Walkman to the beautiful, taped harmony of Vine Song, which Tish had bought for me. The next day, I sailed through the test as I was so spiritually at peace, and fortunately, I could truthfully answer when asked if I had spoken to anyone about the contract while on leave. Then just before he finished, the inscrutable testing officer asked, "Do you believe in God?" I was so taken by surprise by this unexpected question that I gulped, "Yes." This answer seemed to satisfy him as he immediately wrapped up his equipment and removed all the attachments from various parts of my body.

Now that I had a clean bill of health, Adrian became very friendly and relaxed and took me out on the town that night. Most of the locals I saw were black, and I was surprised that, although their ancestors had come as enslaved people over a century ago, they all had the exact characteristics of their brethren from Africa. I was also amazed at how similar the city felt to Harare, in Zimbabwe.

Back on the island

Later that year, a young Nicaraguan pilot called Frank joined our team, although we still took two of our pilots, they asked us to let him fly as a co-pilot. During one of these flights, I was the spare pilot, and as the weather conditions were very turbulent, and after completing the drop, I strung up a hammock in the cabin to relax. At one stage, as the aircraft droned noisily in and out of thunderstorms, the hammock began to swing violently from side to side until a particularly severe jolt broke the knot, and I crashed onto the cold cabin floor.

On another night, the ops room tasked Phil and me to drop propaganda leaflets over Managua, the capital city of Nicaragua. Although we took turns to be the aircraft commander, on this flight, Phil was in the co-pilot's seat. The rainy season had set in, and we did our take-off run during a heavy downpour along a rain-soaked runway, which slowed the aircraft's acceleration. We struggled to get airborne and barely made it by the end of the runway. Our primary task was to drop Special Forces close to Managua, but I suspect that, without realising how heavy paper can be, no one had bothered to weigh the boxes packed full of the leaflets, which had been added as an afterthought. It rained nearly throughout the whole flight, and once we got overhead Managua, we were in the cloud, and the rain was so heavy that we could not see any city lights. So, we blindly scattered the pamphlets, which could have arrived sodden and ineligible.

On the same flight, we then went on to drop some special force personnel in an area of clear weather on the shoreline of Lake Managua from 800 feet AGL. After the drop, Phil was at the controls as we climbed back to our cruising level for the return trip, when I mentioned to him that I had an intense cramp in my stomach. Just as I had finished saying that, and in the same breath and with no change of expression, I continued, "Could you turn right?" To which he answered, a bit puzzled, "What for?" I casually replied, "Because we're being shot at!" I had noticed a curve of red tracer shooting up towards us from the left! This incident then made me realise just how blasé we had become after the number of times we had experienced small arms fire directed at us.

Halfway through the contract, as we had become so comfortable in each other's company, Mark and I volunteered to fly permanently together as a crew. On these trips, we would spend many hours sitting side by side in a dimly lit cockpit. After a while, we lost all inhibitions, and began to chat like two people sitting on a sofa and regale each other with intimate and amusing stories from our past. Before a drop, it was all concentration and business. But once we had carried it out and left the area, a flood of relief would sweep over us, especially after we crossed the Nicaraguan border. We were then able to relax and tell each other stories on the return leg. Some of Mark's antics and situations amused me. Meanwhile, he seemed to find me somewhat quaint and old-fashioned. One night I asked him to promise not to laugh if I told him a particular story. Once he assured me that he would not laugh, I went on to relate to him about my first date with Tish when I was twenty-one, where I had taken her to the eight o'clock show of The Sound of Music. Among all the 'old fogies' sitting around us in the cinema, the first thing to embarrass me was when, in a preview for Meet Me in Las Vegas, Tish let out a wild scream when Elvis Presley appeared unexpectedly on the screen. Later, during the main movie, Captain von Trapp came in to hear his children singing 'Edelweiss'. Their singing completely contravened his instruction to Maria, his new governess, and I expected him to explode in a fit of rage. But, when he joined in their song, it took me so off guard and moved me so much, that tears ran uncontrollably down my cheeks to plop heavily into my lap. Not wanting anyone to know this, I did not wipe them away or bring out my handkerchief or sniff. As I was nearing the end of my story,

Mark reneged on his promise and doubled over, clutching his sides, with his head between his knees, and collapsed into uncontrollable laughter.

The friendship that we shared made my sacrifice away from Tish more bearable. But when Mark learnt that his wife Kate was expecting their first child, he decided to cut his contract short and resign. Feeling I could not bear to continue the operation without his being there, I decided to hand in my notice as well. Following this, Adrian made us an offer: a complete U-turn from their initial policy and said our wives could now come and live in the Cayman Islands. They also wanted to extend our contracts and if we stayed on for another year, they would also throw in a Green Card. But, by then, having made as much money as I would have in twenty years on my previous salary with Court Helicopters, I felt that I had pushed my luck far enough and would like to have a chance to enjoy the fruits of the sacrifice Tish and I had endured. Since that time, I have made my fair share of silly mistakes, but I have also enjoyed the financial freedom that this contract gave us.

30 BOP AIR – START TO AIRLINE FLYING

Bandeirante 110-passenger aircraft

On 16 March 1962, Ian Bond joined as a pilot cadet in the Royal Rhodesian Air Force on No. 16 PTC. This was the same day, we had our Passing out Parade, where I qualified to receive my Air Force Wings.

Ian went on to prove himself a natural pilot and a brilliant sportsman. Apart from playing rugby, squash, and basketball for Rhodesia, he also excelled in cricket and golf. Now, he was instrumental in my becoming an airline pilot.

When the Federation of Rhodesia and Nyasaland broke up at the end of 1963, I joined the Fleet Air Arm and Ian joined the SAAF (South African Air Force), where he soon played rugby for Northern Transvaal, and hoped to go on to be selected for the Springbok team.

When Ian Smith declared independence from Britain on the 11th of November 1965, I returned to Rhodesia after two years in the Fleet Air Arm. Ian later left the SAAF and returned to Rhodesia to become a rep for Rothmans Cigarettes while continuing his international rugby career. Soon after, I drove up alongside him, at a set of traffic lights in Salisbury. Despite never having bought a packet of cigarettes in my life, he thrust a pack of Rothmans through the car window with a friendly smile and a wave.

Later in 1980, when I worked for Court Helicopters in Cape Town, I received a call from him. He had just been appointed as the general manager for Air Cape (which people jokingly referred to as *"Air Escape"*) and invited us to join him and Pat for dinner in Rondebosch. At that time, they were staying in a guest house just before starting his appointment. Later, they moved into a double-story house not far from where we lived in Pinelands and soon joined our bible study group. Our three children went to the same schools and the Friday night youth group meetings. Tish and Pat soon became close friends, and both donated their time to the Christian coffee shop at the Howard Shopping Centre in Pinelands.

Although we lived separate lives, Ian and Pat were always ready to help us through difficult times, mainly brought on by my 'midlife' crisis when I decided to quit flying. The timing could not have been worse as it coincided with our eldest son becoming ill and having to be admitted to hospital for an extended time. I then went for a period without a steady income, with no medical aid, and our car breaking down.

Fortunately, one of our friends Ruth Brown, kindly took Tish to the hospital each day to visit our son. Another friend, Phil Rodger, lent me her car for six weeks, which was the time needed by a third friend, who trained aspiring motor mechanics, to work on our VW Passat at no cost. But it was still a real struggle, until I started selling burglar alarms and later became an estate agent, which allowed me to get back on my feet.

Without realising it, we had forgotten to cancel the

stop order from our bank account, which continued to pay our tithes to Pinelands Presbyterian Church. However, each month, an envelope arrived in our letterbox with enough money to carry us through the month. We were extremely grateful for this, and although we were never sure, I always suspected that these generous donations came from Pat and Ian.

Later, when Namibia gained independence, Air Namibia, a subsidiary of Air Cape, broke away. Their general manager returned to Cape Town, but although I was unsure about the ensuing internal politics, it culminated in Ian losing his position as general manager and looking for another job.

We then heard that he had bumped into Rowan Cronje, a former member of Ian Smith's RF party in Rhodesia, who Ian and Pat got to know when their children went to the same school in Mount Pleasant, Harare. But after Zimbabwe gained independence, Rowan moved to South Africa, where he eventually became appointed as the Minister of Transport and Defence in the newly formed independent homeland of Bophuthatswana in President Lucas Mangope's Government. By the time Rowan and Ian bumped into each other, he was looking for a managing director for Bop Air, the newly established state airliner, and immediately offered Ian the position.

Ian and Pat spent a few happy years in Mafikeng, where they lived in an elegant and spacious company house, which was situated in an elite part of the town and only a few kilometres from his office.

Fortunately, after they left Cape Town, there was a boom in the property market, and as it suited my personality, I became a successful estate agent. At the same time, Tish and Pat continued to correspond.

After two years without a break or holiday, I became so exhausted that when I saw an advert in the Sunday Times advertising a helicopter pilot position in Namibia, I decided to apply. By then, I had come to my senses, and yearned for a more structured life, in the strictly regulated flying world that allocated time off. Once I decided to return to flying, I also phoned some of my friends in Court Helicopters.

After calling Barry Roberts, I applied immediately when he mentioned that Court Helicopters were looking for a pilot. In no time, I was re-employed by them, and although I returned chastened and wiser, I was sadly a bit more cynical about life.

However, after a short time, my life turned completely around. The company first sent me on a contract to the beautiful island of Mauritius for nearly three months. I then went on a lucrative contract, where I worked for the CIA, which entailed flying a fixed-wing aircraft from Swan Island in the Caribbean to carry out supply drops at night to the 'Contras' in Nicaragua.

At the end of this contract, I returned to Cape Town to work for Court Helicopters again. When I signed up for the third time, the financial manager sarcastically welcomed me with, "You are just like a bad penny which keeps turning up!"

After flying for Court Helicopters for a few months, Tish received a phone call from Pat Bond to say Ian was looking for a captain to fly the Embraer Bandeirante aircraft for Bop Air, and asked if I would be interested.

This proposition came just when I was ready to settle down in Cape Town, and I was not keen to move to Mafikeng. From what I knew, it seemed a bit of a Godforsaken place on the edge of the Kalahari Desert. So, I asked Tish to tell Pat I was grateful for Ian's considering me, but I was pretty happy where I was. However, Pat was not so easily deterred and promptly phoned back to offer us free tickets to fly from Cape Town to Mmabatho, to spend a few days as their guests, and see how we liked Mafikeng.

The first thing that struck me was how similar it was to older parts Harare and Bulawayo, with the layout, style of houses, and avenues of jacaranda trees. Of course, Tish immediately fell in love with the place, as she had never really settled in Cape Town, which she thought never seemed like part of Africa. Pat and Ian also went out of their way to make us feel welcome in their gracious company home.

During our visit to Mafikeng, the Operations Manager, Johan Borstlap, who was a short, stocky, brisk, and busy man, interviewed me. I suppose the interview was a formality, as they offered me a job despite my not having a South African fixed-wing commercial licence. The offer also undertook to give me a conversion onto the Bandeirante, a twin-turboprop, eighteen-seat aircraft manufactured in Brazil.

A month later, at the age of forty-six, I started flying as a co-pilot for Bop Air while I started studying for the ATPL exams, which would qualify me to fly as a captain on fixed wing aircraft. Fortunately, I had spent the previous year flying the Casa 212 turboprop aircraft in the Caribbean. So, when I did my conversion onto the Bandeirante with the Chief Training Captain, Mike Kemp, it was most encouraging when he jokingly said he was sure I had a Bandeirante in my backyard.

For the first week in Mmabatho, I stayed at the Molopo Sun. Here I met up with two other newly recruited pilots, Johan Maree and Graham Gush, who would go on to operate out of Jan Smuts. The Ops Manager Johan Borstlap gave us the technical lectures, and after a quick conversion with Mike Kemp, I flew as a first officer for a few months. Bop Air then sent me on a two-and-a-half-month course at Rand Airport near Germiston. While there, I stayed at a boarding house just outside the airport security fence and within walking distance of the classroom.

Over the next few months, I attended lectures every

night, and then continued to study an extra five hours each day in my room. I also made good friends with other students who stayed at the same boarding house. It was convenient to pop into each other's rooms whenever we were stuck on a topic. To motivate myself, I used a stopwatch to keep a strict log of the time I spent studying. Then, whenever I had to stop for any reason, I would pause timing myself. Once I put my mind to it, I could soon go for more extended periods between breaks. Discounting the time spent in the classroom each evening, in just three months I managed to accumulate 400 hours studying. However, when I came to write the exams, I narrowly missed passing all eight subjects on my first attempt. I flunked the plotting exam by just a few per cent and had to then wait a further three months before the Department of Civil Aviation scheduled the next exams. This was such a contrast to the multi choice exam that I had taken at Charlotte in North Carolina, two years ago. There, I had studied for twenty-five hours over two days, and still managed to pass the American ATP exam on my first attempt.

One day at breakfast, we heard a distinctive loud thud to the east of us. That evening, when we arrived in the classroom, we learnt that Bop Air's early morning Bandeirante flight to Jan Smuts, had crashed with the Bop Air Chief Pilot, Geoff Neil, flying as captain. They had been on the final approach to Runway 03 Left at Jan Smuts Airport, when it exploded in mid-air, and crashed into a factory in the industrial area of Wadeville, not far from Rand Airport, where everyone on board was killed. Later, the investigators discovered a volatile liquid in one passenger's briefcase had exploded and caused the crash.

Over the next few months, while waiting for my next opportunity to re-write the plotting exam, I continued to fly as a co-pilot on the daily Bandeirante flight from Mmabatho to Jan Smuts. I would then spend the whole day at the Airport Holiday Inn, where I was free to study. At the end of each day, one hour before the return flight, I would report back to the terminal building's operations room. At that stage, there were no weekend flights, so we settled into a very relaxed routine, flying only five days a week.

Although it was not permitted to copy any of the exam papers, which had to be handed back to the invigilator, one of the ill-prepared candidates managed to make a copy of the whole plotting paper. I was fortunate enough to get hold of these questions, which he had liberally distributed. Having the list of questions was a real bonus, as I could never understand why I had failed this exam, despite it being the subject for which I felt most prepared. However, this question paper was of little use, without knowing which multiple-choice answers were correct.

I felt confident that I had followed the proper procedures and wanted to clarify where I had gone wrong. So, to get an answer, one of the other students and I took a trip to the Department of Civil Aviation in Pretoria, to speak to Mr McGlashlin, who had set and marked the paper.

As we sat on the other side of his desk, I thought he looked like a Battle of Britain pilot with his big bushy moustache. I had hoped he would be willing to explain where I had gone wrong but found him unsympathetic to this request. He simply waffled on, in a most unhelpful and frustrating manner, with my paper, lying on his desk in front of him. As Philip spoke to him about the exam he had failed, I noticed on the desk in front of me a sheet showing the correct list of multiple-choice answers to my plotting paper. I had always been long-sighted, and as soon as I realised what I was looking at, I started to furiously copy the correct letter for each numbered question on a notepad under the desk. Fortunately, Philip kept him talking, as he was alert enough to realise, I was on to something. Then, when I was halfway through scribbling down the letters, Mr McGlashlin's boss called him from his office, and before leaving, he covered the paper. Thankfully, he again exposed it when he returned, and I managed to copy the entire list. Once I copied everything I wanted, I quickly suggested to Philip that we would not be able to get any further assistance and excused ourselves. As we rounded the corner of the long corridor in the austere government building, Philip asked, "What on earth were you writing down?" By then, I could hardly contain my excitement and replied, "You will never believe it, but I got just what I wanted!"

Though, without a copy of the actual exam paper, the list of correct multiple-choice letters would have been of no value. Not having this paper was probably why the examiner was not overly concerned if I saw the master sheet, he used to mark the papers. Now having the correct answers, I immediately realised where I had gone wrong when I went through the exam paper. It had nothing to do with my technique or knowledge. But instead, I had chosen to plot the entire route before choosing from the multiple-choice answers. What I should rather have done, was to plot just as far as the next question. I should then set off from the position closest to where I had reached at that stage. No two people will arrive at the precise location when drawing a pencil line on the plotting chart. So minor discrepancies will always creep in when using navigation rulers to measure distances, and protractors to measure the angles. To avoid the slow but continual divergence from his plot after each question, I needed to start afresh from the examiner's choice of positions. The time allocated for this exam was three hours, but after a while I could complete the whole section for the plot, in fifty minutes.

By the time the next exam came around, I was confident that I would be able to sail through. Luckily, before the exam, I bought a second-hand calculator designed to work out the weight and balance C of G when loading an aircraft. Although it was part of the syllabus, the previous exam had no questions on this subject, so I did not practice how to operate this little machine. On the re-write, I only managed to complete the plot with five minutes to spare. This left the remaining questions to account for 30% of the total marks. To guarantee a pass for the exam, I would have to achieve 100% of what I had managed to complete. I knew there was insufficient time to do the calculations in longhand, and in sheer panic, I realised, that the only chance I had to finish in time would be by using the calculator I had just bought. This little calculator was legal and acceptable, as there was no requirement to produce handwritten calculations, except, I had not spent much time learning my way around it.

Despite telling us they would confine their questions to the 747 (Jumbo Jet), this exam paper used the out-dated Boeing 707 for the weight and balance questions. Fortunately, the answers were progressive, leading from one question to the next. With time rapidly running out, I could hardly breathe. Then I saw one solution after the other pop-up on the small screen corresponding to the multiple-choice answers. As the invigilator called out, "Pens down!" the surge of adrenaline left my hands trembling as I entered a cross against the final solution. As we stood up to hand in our papers, I felt ready for a stiff brandy to settle my nerves and celebrate my miraculous good fortune. Of course, this time, I sailed through.

But I feel the extra time flying as a co-pilot on the Bandeirante was invaluable, as it allowed me the opportunity to fly with several highly experienced fixed-wing captains. A few of them were retired training captains who had spent many years flying for SAA (South African Airways). Mike Kemp, Cyril Rogers, and Willie Coertze displayed professional cockpit resource management skills.

They had introduced the 'monitored approach' principle in which the co-pilot was the flying pilot in poor weather conditions. The captain would monitor the instruments and focus his attention outside the cockpit to look for the runway in the latter stage. When he could see the lead-in lights and sufficient runway lights, he would take over and land the aircraft. SAA and other airlines worldwide, employed this method, which contrasted with the early days when co-pilots did not dare question the captains, who would fly the letdown and then must look up at the last minute, refocus their attention outside the cockpit, and carry out the landing.

Once I became a captain, I progressed from the Bandeirante to its bigger brother, the Brasilia, and then the DC-9, the MD 80, and the Boeing 727. In between, I flew Citation II and Gulfstream III in a corporate capacity and made some interesting overseas trips which I later describe.

I will always be grateful to Pat and Ian Bond for persuading us to join Bop Air, which eventually expanded to become Sun Air, until SAA finally bought it out, and then liquidated it.

31 CAPTAIN, ONE THING I CAN'T DO IS FLY

Cessna Citation II

Dino Paneris and I were on standby one Saturday when the ops room called and requested us to fly eight passengers to Pilansberg Airport, which served Sun City, in the twin jet Citation II. One of the passengers was Minister Radebe, the Deputy President of Bophuthatswana.

We took off with clear blue skies as far as the eye could see and had a relaxed and uneventful flight to Pilanesberg. After twenty minutes, we descended over an ostrich farm and a low scrub of thorn trees to touch down on the newly tarred runway. We then taxied onto the parking area in front of the attractively thatched and ethnically designed terminal building. After informing us that they were going to lunch and would be back at the airport at 1500 hours, two black Mercedes whisked our illustrious passengers off. We took the transport to the Sun City hotel complex, where we ate our in-flight snacks and cold drinks while relaxing on reclining chairs around the sparkling pool at The Cascades Hotel. It was such a glorious afternoon, with the gentle sun warming the bathers, that it never entered our heads to phone the met office and check the weather for our return flight. Eventually, we dragged ourselves back to the airport just before 1500 hours. We did the pre-flight checks and were ready, with the co-pilot strapped in his seat, and awaited the arrival of our passengers. When they eventually arrived three hours later, I stood at the cabin door to welcome them on board.

Our arrival at the airport to have the aircraft ready in time was always guessing game. We would see how finely we could cut things and still have everything prepared before our passengers arrived. Our operations manual required us to be at the aircraft a whole hour before the passengers were due, but as usual with this grade of clientele, getting us to be ready up to three hours early had become the norm. I had previously flown Minister Radebe in various aircraft and, although I always stood respectfully to the side of the steps to welcome him on board, he had never once acknowledged my presence.

I realised it must be a racial thing, which amused me. I think this minister wished to make a point in this new black homeland, which had been established as part of the Apartheid policy, that I was now a second-class citizen. Other dignitaries felt that the pilots were more than baggage boys anyway. At times we had to carry cumbersome luggage to the apparent delight of the passengers who relished in the opportunity to lord it over these 'white peasants' while their security guards stood by idly, without offering a helping hand. Fortunately, I took no offence. But sometimes, when the first officer passed me the luggage from the back of the aircraft, I would amuse myself by letting it pile up at the top of the stairs and then gently stretch back the toe of my shoe to send a fancy elephant-hide briefcase flying down the stairs, onto the hard-standing. I then waited for one of the security guards to get an admonishing blast for not taking more care.

On this day, the passengers finally arrived just before the last light. Once again, we took off in clear blue skies as we set course for Mmabatho. Almost as a formality, we turned on the weather radar. With the flight a short twenty-five-minute hop, we certainly did not expect the weather conditions to be much different from what we had experienced during the incoming flight and enjoyed while we sat basking around the pool. But as the screen lit up, we were shocked to see a big red blob centred at ninety nautical miles, which was on the direct path and exact distance to our destination. Apart from that, the radar was clear of any other weather in the whole region. The bright red indicated a raging storm, and we prepared for the worst. We decided to attempt a landing at Mmabatho, knowing that we could still divert to either Wonderboom or Lanseria airport, if we could not get in there. According to Area Control, both alternative airports were open and in clear weather.

Typically, during thunderstorm activity, the ADF needle, which we had tuned to the Mmabatho NDB, swung around erratically. There was no ILS facility to assist our letdown approach to the runway at that time. The only navigational aid which continued to function was the VOR beacon. Therefore, we were committed to a VOR letdown, with the VLF navigational system as a backup, to confirm that we were directly over the beacon. From there, we would follow the VOR letdown procedure.

It was dark outside by the time we reached fifteen nautical miles south of the field, and, although we braced ourselves for the ordeal, it still took us by complete surprise when we hit the edge of the towering cumulonimbus. The turbulence was nothing like anything I had previously experienced, and the aircraft bucked so violently that my head kept smashing against the overhead bulkhead. Dino helped me tighten my lap straps, while I fought to hold on to the control column which whipped about crazily. Heavy raindrops drummed against the top of the fuselage and windscreen so loudly that we had to shout to hear each other speak. While keeping the aircraft flying under control, we noticed thick ice accumulating on the wings' black leading edge. We had to activate the de-icing rubber boots to break it up before it seriously affected the aircraft's flight performance. If neglected, it could quickly spoil the smooth flow of air over the wings and drastically reduce the lift and aircraft's ability to stay aloft.

We had switched on the heating elements, sandwiched between layers of the forward windshield, for de-icing purposes before take-off. We had to activate the engine intakes heaters to avoid the danger of ice build-up and reduction to our engine performance.

While tossed about like a cork on a turbulent sea as we battled to carry out approach checks and do the let-down briefing, we heard loud, eerie wailing from the cabin behind us during our struggle to keep abreast of the situation. Our erstwhile aloof passengers were obviously in the grip of terror. With enough on our plate and doing our best to stay calm and professional, the last of our concerns was to reassure our panic-stricken passengers by making an announcement.

We fought to control the aircraft with the rumble of thunder and blinding flashes of lightning all about us on the let-down. It was with relief that the runway lights appeared through the downpour just as we approached the break-off altitude. However, the landing lights merely reflected off the rain. A fierce crosswind drove the rain slanting sideways across the runway to add to our woes.

The runway was awash, and we fought to avoid aquaplaning or sliding off the right-hand side of the tarred surface. I had to crab the aircraft by pointing its nose into the crosswind to keep on the approach's centre line. Just before touchdown, I yawed the plane straight at the last moment while dropping the left wing to land on the port wheel.

When I finally brought the aircraft under control, we cleared onto the taxiway at the end of runway 04. By then, the adrenaline could kick in as I took a deep breath and was finally able to relax.

We must have been in the eye of the storm because, as the passengers disembarked, the rain ceased just long enough to allow them to exit the aircraft and climb into their waiting limousines.

Minister Radebe, who had never previously returned my greetings, approached me as he reached the ground. Taking my hand in both of his and wringing it fervently, he said with a trembling voice, "Captain, thank you, thank you so much for getting us down safely. There is one thing I cannot do – I don't know how to fly!"

Before we left Mmabatho earlier that day, we had parked our cars in the hangar, and I hoped we would be able to taxi straight in, to avoid being drenched in the torrential downpour. When our passengers had departed, we climbed back into the cockpit, just before the rain began beating down heavily again. After starting up, I taxied the aircraft right up to the hangar doors which were closed against the weather. Unfortunately, the rain drummed so heavily as it thrashed against the metal-clad hangar door that it drowned out the noise of the jet engines, and the hangar staff remained oblivious to our presence. We were obliged to sit in frustration for the next fifteen minutes before the storm slowly moved off to the east and allowed us to dash for the hangar door, which we opened just wide enough to slip through.

Sadly, despite surviving this terrifying experience a short time later, Minister Radebe was tragically killed in a car accident.

32 YES, BUT YOU DON'T SEEM TO DO VERY MUCH

Embraer 120 Brasilia Aircraft

One Sunday, I was the captain of a flight between Durban and Pilansberg, the airfield that serves Sun City. At that time, Sun City was a gambling resort in Bophuthatswana and close enough to attract punters and other holidaymakers from Johannesburg and Pretoria. It was a lavish complex of hotels, a golf course, and a game park, established in the bowl of a colossal meteorite that had struck the ground about three hundred million years ago. Gambling was still illegal under the strict gambling laws imposed on South Africa in the Apartheid era.

I captained our scheduled flight, which started at Mmabatho Airport in the morning and flew the first leg to Johannesburg International Airport. From there, we uplifted passengers for Durban. Many of these passengers had been brought from Pilanesburg by the Hawker Siddeley HS 748 (a lumbering 48-seat aircraft called the *Hawker Slowly*. After a short stay on the ground in Johannesburg, we continued to Durban to collect passengers, fly them directly to Pilanesburg and then return to Mmabatho.

At that stage, I was the Bop Air Operations Manager, and my office was next door to the hangar at Mmabatho Airport. To have a break from my office during teatime and lunch, I would often join the technical staff in their crew room in the hangar building.

There has always been a friendly rivalry between the technical staff and pilots. However, I have forever laughed it off as pure jealousy. I would tell them that I was not responsible for their wrong choice to go into aviation's technical side. I could not help it if they chose a vocation, which entailed getting oil and grease under their fingernails. When we wanted to dig at them, we would call them "Grease monkeys", but they had other names for us, which are unprintable.

Sour grapes convinced aircraft technicians that we did not warrant our supposedly inflated salaries and time off between schedules. The stipulated rest periods, called Flight and Duty Limitations, were out of our hands. The Civil Aviation Authority set the strictly regulated restrictions as a safety precaution to ensure pilots had sufficient rest.

On this day, when we landed at Durban, one of our technical staff, John Wentworth, and his family came on board. They had taken some of their leave in Port Elizabeth and were now returning to Mafikeng. They had arranged rebated travel to Durban, where they joined our scheduled flight in the twin turboprop Brasilia aircraft.

It was a beautiful clear day, so we invited two of John's elder children to take turns in joining us in the cockpit. First, we had Nicholas, who was about seven years old. Before start-up, we settled him on the 'jump seat,' just behind and between the captain and co-pilot seats, where he was able to enjoy the take-off.

Giving him the spare headset, we explained a few

things to him as we went along. He was also delighted to hear communication between us and the air traffic controllers.

That day, we quickly settled on the track for Pilanesberg, via the preselected waypoints from the navigational log, and entered them into the GPS. I was the flying pilot for the leg while the co-pilot did all the radio calls, monitored the flight log, and set up the frequency for the various VOR and NDB navigation ground beacons.

I had engaged the autopilot, which kept the aircraft heading towards NDB and VOR beacons that we could see on the radar screen for each flight section. The 'glass instrument' screen depicted the track as a line between each waypoint. Simultaneously, the autopilot steered our Brasilia along this route that we had pre-programmed into the system.

Once we settled at our cruise level, the flight progressed smoothly, and the scenery drifted gently below us. Over the cabin speakers, I pointed out various features to the passengers as we passed Mooi River, Giant's Castle, and Cathedral Peak along the edge of Drakensberg Mountain. We could also see Ladysmith and the Vaal Dam, which I mentioned to them.

Halfway through the flight, I asked Nicholas if he thought his 12-year-old sister Sarah might like a turn so one of the air hostesses could settle her into the jump seat before the landing. He went back to call her.

Sarah was a serene young lady with maturity and composure way beyond her years. When she came up, I was busy reading the Sunday newspaper while the autopilot took care of the flying and navigation. It automatically changed course onto the next heading as we went overhead at each waypoint along the route. On this leg, the co-pilot monitored the flight log and made a radio call to report entering the Control Boundary for the Johannesburg Area. He then changed frequencies to check in with the various air traffic controllers along the way.

Once we passed the built-up area around Johannesburg, I asked Sarah if she liked my job. She answered hesitantly, "Yes, but you don't seem to do very much."

The next day at work, I related this to the aircraft technicians, who call themselves 'Engineers'. This story just cracked them up, and after that, they often told the story to the delight of other engineers in the industry.

I suppose small things will always amuse small minds!

33 NIGHT FLIGHT TO THABANCHU AIRPORT

Bop Air Citation II

One Sunday in 1991, I was on home standby for any Bop Air corporate Citation II aircraft request. At the time, we lived on a small farm, 19 km to the east of Mafikeng, and had called out a local veterinary doctor, Deidre, to inject our horses against horse sickness. When she finished administering the injections, we invited her to join us for lunch in our rambling five-bedroom house, part of which the first owners had built before the Boer War. It was a charming house with two bathrooms, knotty pine ceilings, Oregon pine door frames, skirting boards, and oak doors. As we sat at the table, I received a call from Aubrey, the duty ops assistant that day. He told me he had just received a request for the Citation to fly to Wonderboom Airport near Pretoria and take the Minister of Education, to Thabanchu, near Bloemfontein. At that time, Thabanchu was part of Bophuthatswana's independent homeland's fragmented and dispersed sections.

As the Diedre's brother was a pilot with SAA, she was keen to come for the ride. I told her that we would welcome her on board if she got to the Mmabatho Airport by the time I was ready to depart. But I explained that once the Minister arrived, there was a chance that he might not agree to her accompanying him on the flight from Pretoria to Thabanchu. She assured me she would be happy to take the chance as, if necessary, a friend would be happy to collect her from Wonderboom and give her a lift to her parent's house in Johannesburg.

After a hurried trip to Wonderboom, we hung around for hours before the Minister finally turned up for his flight to Thabanchu. Once again, I wondered if the minister's secretary had asked us to get there long before the expected departure time to avoid his being delayed or inconvenienced in any way.

By the time he arrived it was already dark, which would make it necessary to involve a night landing at the small airfield in Thabunchu. To get there after dark did not particularly bother me as I had landed there many times before, having flown the Piper Chieftain, King Air 200 and Bandeirante in and out of Thabanchu a few years before, when we operated a daily scheduled flight from Mmabatho.

Those were glorious days when I would change into my running shoes, shorts, and vest to the amusement of the check-in and airport staff, while the co-pilot took my uniform on the hotel transport to Thabanchu Sun Hotel. I would then run uphill on the five-and-a-half-kilometre road to the hotel. Once there, I would shower, have a snack in my room, and then have a short sleep. When I woke up, I would go to the in-house gymnasium for a workout, sauna, and shower before lunch. Later that afternoon, we would fly the schedule back to Mmabatho.

I had never met this Minister before. He was an ordained minister, a large doughy man with a limp handshake and referred to me as, *"My dear"*. He was, however, more than happy for our vet, Dr Deidre, to accompany him on the trip. Deidre was over the moon and graciously attentive to the minister during the flight and served him snacks and cold drinks that the ops staff had stacked in the ice-packed, metal-lined, drawer.

The first half of the flight was uneventful until we noticed the sky on the horizon lighting up with bright flashes of lightning that outlined an ominous thunderstorm, sitting just off to the northeast of the airfield. When we were close enough, we were relieved to see the runway lights shining bright and clear, with the storm raging a few miles to the northeast. Since I had operated at this airfield on many occasions, I knew the area's high-ground layout. The big mountain on the overshoot was the one I used to jog up a few years ago, and I was also familiar with the small range that ran parallel on the left-hand side of the approach path to the airfield, when landing from the west.

I quickly identified this range's position by the red lights of radio masts on the ridge and realised that we would remain clear of this range if I kept south of these lights, while positioning onto the final approach. However, the storm beyond the airfield made it impossible to overfly the airport and carry out a regular circuit pattern.

With the experience I had gained doing night supply drops to the Contras in Nicaragua, I thought nothing of doing a circuit to the west of the runway, while keeping south of the beacons, as I let down to circuit height. It would be perfectly safe if the runway remained visible for our landing approach.

Unfortunately, I failed to consider the co-pilot's discomfort with my non-standard and unconventional procedure, that did not conform to the field's laid-down approach pattern. His unease was exacerbated by being unfamiliar with the layout of the airfield and its surrounding high ground. Having been in similar situations with other captains in my flying career, I should have been more aware that he might have been freaking out and disorientated throughout the whole procedure.

As it was, we landed safely, and then after dispatching the passenger, departed from our landing approach's reciprocal direction. We climbed safely away from the storm, which continued to rage over the mountains to the east.

I did not realise just how uncomfortable the co-pilot had been until the chief pilot asked to speak with me the next day. He informed me that the co-pilot had complained about the unorthodox procedure I had adopted. This co-pilot had joined the SAAF (South African Air Force), where he had only flown small, single-piston-engine aircraft. With no experience on jet aircraft, he was extremely fortunate to have been in the right place at the right time and managed to go straight onto jet aircraft, without first flying any of Bop Air's turboprop aircraft. His limited background and experience made his complaint fair and understandable, especially as it made him feel disorientated.

At that time, I was approaching 50 years old, and it made me realise just how little one's background and experience count; as one grows older, the younger generation sees us as older men. I well remember how arrogant and judgmental I had been towards some of the more senior pilots, when I was coming up through the ranks.

After dropping off the minister, we returned to Mmabatho with Dr Deidre. She had been entirely unaware of the drama and was just thrilled by this unexpected adventure!

34 GULFSTREAM FLIGHT FROM ABIDJAN

Gulfstream III

Bop Air purchased a Gulfstream III from Anglo America in 1992. The company sent Mark Fenton and me to the Savannah, Georgia, Gulfstream factory, for lectures and simulator training. By the middle of October, Bop Air Engineers operating from the Jan Smuts Airport had fully refurbished our new aircraft. Our MD asked the Anglo America aircraft division's Chief Training Captain, Tony, to undertake our flight training and conversion onto the Gulfstream. The company then asked him to join us on our first overseas trip to take President Lucas Mangope of Bophuthatswana to Tel Aviv for four days and then on to Munich for eleven days.

The Minister of Aviation, Mr Rowan Cronje, put us under tremendous stress by committing us to depart on the 11 November 1992. At first, we could not get anywhere near the aircraft while an upholstery company refurbished it with soft, leather seats. The General Manager of Engineering then sent it to be painted into Bop Air livery. Nevertheless, the result was perfect. The colour scheme was predominantly a high gloss white with an impressive leopard's head on the expansive tail fin. To finish it off, they added thin red and blue stripes running the whole length along the fuselage's sides, which gave the aircraft a clean, uncluttered look, and showed off its classic lines.

Our Gulfstream was aesthetically beautiful with its swept-back wings, upturned winglets, sharp pedigree nose, and engines at the rear of the tapered fuselage. The overall appearance suited its sleek design and hinted at its impressive high-speed performance. The Rolls Royce engines produced so much raw thrust that we battled to keep within airspeed restrictions in controlled airspace. At each level off, during stepped climbs, it was also hard to believe how far the throttles had to be retarded to keep the speed down. Even above 45,000 feet, cruising at Mach .85, we still had the throttles just above the idle position.

It was challenging to produce a new handheld checklist to conform to the Bop Air format. Our procedures were foreign to the Anglo-American instructor, who was against any divergence from their methods. The paradigm shift was just too much for him to accept, and he could not feel comfortable with them.

Our modus operandi was introduced to Bop Air by Mike Kemp and Cyril Rogers. These two retired SAA training captains had joined the fledgeling airline of Bop Air just a few years before me. I tried to explain to the instructor that our procedures were in line with all the major airlines worldwide.

The Anglo-American procedures aligned with other corporate jet companies and used a small roll-up checklist above the instrument panel. On the other hand, I found the Anglo-American method somewhat confusing, especially in how they carried out their 'approach checks' in a progressive, stop-start fashion while descending for an approach to land.

I felt our system was more structured where the 'flying pilot' would call, "Approach Checks please", at which the non-flying pilot would then carry out the checks in silence. Once finished, he would respond with, "Approach checks complete." On the approach, the flying pilot would then call for 'Speed brake' and the various stages of 'flap' they would need to maintain the correct descent profile, while reducing speed as required. On the last stage of the final approach, and before commencing the descent on the 'ILS,' we would lower the gear and then select full flap. When we had fully configured the aircraft for the landing, the flying pilot would call, "Landing checklist, please." As the non-flying pilot read out the items, both pilots would double-check that each selection was correct, and finally, the non-flying pilot would end off by calling, "Landing checklist complete!"

We also did what was known as a 'monitored approach' whenever we had to carry out a let-down in marginal weather. During this, the co-pilot would fly while the captain monitored the flight and engine instruments. The captain would also look ahead for outside visual references, such as the lead-in lights to the runway. The co-pilot made the 'callouts' for specific heights, such as "One hundred above", and then he would call, "Decide" at the break-off altitude. At this point, the captain, focusing most of his attention outside the cockpit, would either call, "Continue" or "Go around!" If he called "Continue", the co-pilot would continue to fly the aircraft by reference to the flying instruments, until the captain called, "I have control". At that stage, the captain would take over control of the aircraft for the landing. If he did not have enough outside references to land, he would call, "Go around!" In this case, the co-pilot would carry out an 'overshoot' while still flying on instruments.

Tony was more familiar with the outdated procedure, where the captain approached by flying on the instruments and looking for outside references to carry out the landing himself. However, 'Cockpit Resource Management' considered this technique unsatisfactory, as it took time for the captain, flying on instruments, to readjust his focus outside the aircraft and then still land safely.

This difference between our procedures and checklist philosophy contributed to considerable tension during our training. Tony just wanted to adhere to the methods he had used in the Anglo-American company. He was such a hardened sceptic that it was impossible to convince him there was any merit in the procedures that become accepted by most airlines. His uncompromising attitude put undue strain on what was already an uncomfortable task to get us up and running in time for the trip.

Tony had done numerous overseas flights in this Gulfstream when it belonged to Anglo and knew the plane well. While we gratefully accepted his expertise and experience, we continued to struggle with his inflexibility. He was also more 'old school' and rigid about cockpit protocol, which contrasted with the more relaxed and friendly relationship we enjoyed with our co-pilots. His philosophy was not to encourage them to have any input or opinion. This approach also exacerbated the cockpit's tense atmosphere, and my co-pilot, Mark Fenton, took exception to his condescending attitude.

However, we did our best to maintain our normal relationship in the cockpit while at the same time trying to adapt to the aircraft. Although it was exhilarating, it could quickly get ahead of us with the slightest lapse in concentration. As it was, I always found converting to a new aircraft both physically and emotionally exhausting. However, in this case, the added responsibility of planning the route added more stress. It entailed working out flight-plan logs and obtaining international clearances from the countries we planned to overfly. This task was not easy, as some African countries refused permission to fly through their airspace if the destination was Israel. Some corporate pilots advised me to put Larnaka Airport in Cyprus as our destination to overcome this obstacle. Then once we were halfway across the Mediterranean, we could call for a diversion to Tel Aviv.

The day before we left, I was up until after 2300 hours that night finalising flight navigation logs on a computer program I had acquired explicitly for this task. Using it, I could print out the navigation logs with all the waypoints, distances, headings, and times for each leg. It was so new to me that it took longer than it should have. Still, at the same time, it saved having to measure everything on navigation charts manually. Even so, because of the late night, I felt exhausted the next day. When we settled down in the flight and were overflying Malawi, it finally caught up with me, and a blinding migraine hit me.

When we landed at Jomo Kenyatta Airport in Nairobi, I was in no mood to pay precious American dollars as a bribe for 'handling fees' and spurned their offer. My reluctance to pay handling fees resulted in an entirely non-cooperative attitude from every Kenyan I approached to direct me to the landing fees office. With my head still splitting, I walked from one end of the semi-circular terminal complex to the other. Fortunately, I met a sympathetic Israeli in one of the hangars, who accompanied me to the office outside the terminal building on the other side of the parking lot!

When we arrived in Tel Aviv, the President's security guards offered no assistance to offload the matching oversized leather suitcases. They weighed so much that the task required two pilots to do the job. We then had to lug them to a large trolley at the President's beachfront hotel.

We then checked ourselves into the crew's hotel, which was luxurious enough. I particularly enjoyed selecting Mediterranean treats for breakfast and chose rollmops, smoked tuna, and delicious cheeses.

The next few days, we experienced a taste of Israel as we took guided tours to Jerusalem and Bethlehem. We followed the Via Dolorosa with an Israeli guide who surprised me with his total bias against the possibility that Jesus was the Messiah or could have risen from the dead. He specifically went out of his way to disprove that Jesus could have died in such a short time and tried to convince us that his disciples had taken him away.

I was amazed at how much of Jerusalem's city in Jesus' time was far below ground level and how many different churches there were in such a small area. We also had a chance to visit the Mount of Olives, Gethsemane, and the Wailing Wall.

On our first night, the Bophuthatswana Ambassador and her husband took us to the most vibrant place I had ever experienced. It seemed to be a combination of a pub and a restaurant with rows of solid wooden tables, benches, and a live band. What struck me was how people spontaneously started to dance all by themselves. They would do this in a most liberated fashion while enjoying the wonderful Israeli music. Mapala, our Bop Air Tswana cabin attendant, also jumped onto our table and started dancing in her gyrating African style, which was a bit embarrassing to the rest of the crew!

When we had been in Israel for two days, we took a trip to the Dead Sea and went up the cable car for Masada's tour. I then had a chance to float in the bitter, oily water of the Dead Sea and, following tradition, smeared thick black mud all over my body, which I allowed to dry and crack on my skin before washing it all off with a refreshing shower. As it washed off, the tension finally thawed out of the knotted muscles in my neck. I felt emotionally and spiritually revitalised as we journeyed back to Tel Aviv that evening.

The next day we took off for Munich. When we made radio contact with ATC, they told us that the cloud base was down to 200 feet, the minimum height which allowed us to land there. So, I informed Mark that we would have to carry out a 'monitored approach.' Although this worked out perfectly, and despite neither of us having been to Munich before, our instructor could still not acknowledge this procedure's merit.

After landing, taxiing to the apron, and shutting down the aircraft, Mapala opened the cabin door, automatically releasing the unfolding stairs. As she stood in the open doorway dressed only in the company's soft blue patterned uniform, the freezing temperature immediately hit her. She was a light-skinned, pretty Tswana woman, and it was the first time she had been outside South Africa. It was also the first time she had ever seen snow, and after the warm weather in Israel, this cold weather came as an unexpected and unpleasant shock to her system.

I monitored the cheerful, blond, pink-cheeked German lad as he cleaned out the bilge tank. He was full of beans but spoke no English. Fortunately, he got the gist of what I said when I spoke broken Afrikaans. When I told him we must now "*roer*" the mixture in the toilet, meaning he should 'flush' the toilet to swirl the disinfectant mixture around, he answered, "*Javol, spoel!*" to which we both roared with delight at our success in breaking the communication barrier.

It was my first visit to Germany, and I was impressed by the modern aluminium-framed glass airport terminal building. I was also amazed at how honest the commuters were on the Munich Subway. The commuters bought their tickets from ticketing machines along the stations' sidewalls and then walked straight onto the platform and a carriage unimpeded. One day, we took a tour around the BMW 3 Series factory and were impressed at how efficiently and methodically the assembly line people worked. We were also able to take a trip to the mad Swan King Ludwig II's fairy-tale Neuschwanstein Castle, which had inspired Walt Disney's Sleeping Beauty's palace. It is a breathtaking sight with ornamental turrets, gabled roofs, and balconies perched on top of a sharp cliff ridge. It looked like it came straight out of a fairy tale book and gave the impression of being built like a knight's castle from medieval times, so I was surprised by its relatively modern kitchen until I learnt they had constructed it only 100 years before.

After we toured the castle, we took a trip to a local ski resort in the Bavarian Alps on a ratchet-driven coach up an unbelievably steep climb. Our coach was full of healthy-looking German youths dressed in colourful acrylic ski suits and carrying short snowboards. The ski resort was buzzing with activity, and we enjoyed the atmosphere as we treated ourselves to decadent cakes and sipped rich, steaming, hot glühwein.

We enjoyed Munich, although the residents did not seem overly friendly towards English-speaking people. Very few of the older folk would speak English in the shops, but this became somewhat understandable after discovering that the Allies had subjected the Bavarian industrial areas to severe bombardment during the Second World War. Our visit was just before Christmas, and the pavements were slippery, with the solid ice proving tricky as we walked around. But the street and shop decorations gave it a festive feel. I enjoyed walking, but I had to wear my 'Long Johns.' Dressed in my thermal underwear was exhilarating while outside but became unbearably hot whenever I entered any shops.

Whilst strolling and window-shopping, we passed many small shops selling knuckles of pork rotating on rotisseries in the windows. Then in the evenings, we visited one of the large beer halls. There, we soaked up the atmosphere enhanced by the *'oompa'* bands and watched barrel-like men delving into huge pork knuckles and *sauerkraut*. One evening I thought I would try a half-sized one but struggled to finish it. I realised why, with it being so fatty, the acidic *sauerkraut* was necessary. I had previously heard how the waitresses could carry many litre beer mugs in each hand but had always imagined they would have to be big blousy girls. What was more impressive was how strong even the young petite ones were.

We left Munich for Abidjan on the Ivory Coast on a crisp, clear winter evening. We set course, overflying the snow-clad Italian Alps, and the Mediterranean Sea over Sardinia and continued towards Africa's north coast. While Mark spoke to the air traffic controllers, I had a chance to enjoy the sparkling necklace of lights along the Italian Riviera and the inky black sea dotted with the pinprick lights of vessels.

In no time, we were over Algeria and then the vast emptiness of the Sahara. After serving a sumptuous dinner over the Mediterranean, Mapala and the passengers settled down to sleep. Despite asking her to carry out periodic checks in the cockpit to ensure we were still awake; we never saw her again until we landed nearly six hours later.

The designers made every effort to make the pilot seats as comfortable as possible by including fitted sheepskin covers and adjustable pads in the lumbar region and below the thighs to relieve sitting for hours. Nonetheless, they eventually became rock-hard. Therefore, we periodically got out of our seats for some refreshment, stretched our legs, or went to the toilet. However, after things had settled down, I opened the cockpit door and was surprised to discover that the cabin was in total darkness, and our air hostess and passengers were all covered with blankets and fast asleep.

The flight was nearly seven hours long, and we spent over three hours flying over the Sahara Desert while the Gulfstream covered eight nautical miles per minute. When we finally reached the top of the descent, we made a call over the passenger loudspeaker to warn Mapala that we were twenty minutes out. The announcement should have given her enough time to prepare the cabin, wake up the passengers, and secure their seat belts for landing.

Our arrival at Abidjan was in the middle of the night, and as I was the flying pilot, I set up a long slow descent. I misjudged it and ended up too 'hot and high' for a straight-in approach, to my embarrassment. So, I had to make a wide circuit as we swept overhead the sleeping city lights, nestled on the Atlantic coastline.

After all that, I was even more alarmed when we shut down the aircraft in the parking area and eased out of my seat. I discovered everyone, including our cabin attendant, still fast asleep in the bunks and folded back seats. I was quite amazed, as this was so out of keeping with the Bop Air cabin staff's professional standard. When I pointed it out to Mapala, she was surprisingly unrepentant or apologetic. She said she did not see why she should disturb the sleeping passengers and prepare the aircraft for the landing.

I was further surprised when the cleaners came on board. The passengers had still not budged, and the President slept soundly on his bunk at the rear of the aircraft while the cleaners brushed past him.

Mark filed the flight plan while I paid the landing and handling fees. I also monitored the airport staff cleaning the bilge tank, which they accomplished in the most primitive manner by draining it into handheld buckets. They then carried them away, sloshing all over the hard standing. In the meantime, Tony monitored the refuelling and set up the INS (Inertial Navigation System), which we found pernickety, and which I was still not confident about. While we entered our location's latitude and longitude coordinates, the aircraft had to be kept steady without any jarring movements. After that, once the plane started to move, the instrument system would sense the acceleration to update our position. We would have to enter the 'waypoints' for the trip, and during flight, we could update the instrument to the exact coordinates when we passed overhead a known position.

Our route from Abidjan took us over the sea for a few hours to a waypoint on the Jeppesen aeronautical chart called 'Ildir'. This waypoint was right on the Greenwich Meridian and one hundred nautical miles off the west coast of Africa and in line with the border between Angola and Namibia. From there, we planned to turn and head directly for Mmabatho Airport and pass just to the north of Windhoek.

We took off just after 0200 Zulu (Standard Greenwich Time, or Universal Standard Time, used in all aviation worldwide). We quickly headed out over the blackness of the Atlantic Ocean, with the lights of Abidjan rapidly disappearing behind us. As we climbed into the starlit sky, the INS unexpectedly called for a heading of 170 degrees which was right of our flight plan, heading for still air conditions. Although we had the radar sweeping on the 'Map' mode for the next few hours, it did not show Africa's coastline, to help provide further assistance.

The unexpected heading immediately gave me an uneasy sense of foreboding which I mentioned to Tony. However, he simply shrugged it off with his now familiar, longsuffering disdain. Regardless, I could not shake the lingering feeling of unease, especially when I remembered that fateful night less than three years ago when I was employed by the CIA and had to ditch in the Caribbean Sea.

Now, as the non-flying pilot, for this leg, it was my responsibility to keep abreast of the navigation and to fill in the flight log at each waypoint. I was also responsible for the radio calls. But, as we were over international waters, this only required the occasional general transmission over the standard VHF frequency to alert other aircraft of our call sign, route, direction (north or south), flight level, position and estimate for the next waypoint.

The cockpit became incredibly quiet as fatigue set in after an already long flight, and both Mark and Tony dozed off. Since I had sat in the sheepskin-covered seat for so long, occasionally, I had to wriggle around to ease the numbness and ache in my back.

Not surprisingly, the VOR and DME beacons along the West Coast, which would have made it possible to pinpoint our position, were all out of action. We had found the same lack of navigational facilities over most of the African countries we had over-flown. Nevertheless, I dutifully continued to transmit.

position reports on the HF frequencies. I listened to the loud crackling static from far off and unknown places in distorted 'Donald Duck' squelchy speech and indecipherable accents. Despite having no navigational facilities in operation, the radio operators went to great lengths to take note of the aircraft registration and name of the operating company, with the sole purpose of submitting an account for overflying their airspace. It is no wonder that as no navigational aids were available, some operators forsook the radio calls to avoid having to pay the exorbitant fees for only flying through their airspace.

As we flew on, I tried to quell the feeling that we were diverging from our planned flight path and continually tried to get a fix from one or more of the many VOR/DME beacons depicted on the aeronautical chart for each country we passed. I could still not understand how there could be such a discrepancy between our calculated 'headings' between the waypoints and what the INS said we should steer, or how we could have been so far off-course from the moment we got airborne.

As I searched the heavens for various known clusters,

I could see The Southern Cross and its Pointers. For the first time, I even thought I could make out what must be the 'Crab,' the 'Plough', the 'Scorpion', and the 'Taurus bull'. Suddenly, as I gazed at the night sky, I saw a glowing ball of fire shimmering as it headed straight towards us. As I sat transfixed, it rapidly enlarged. It all happened so quickly that I never had enough time to wake the other sleeping crew members. I was mesmerised and convinced that we were on a direct collision course. I was unsure what evasive action I should take to avoid what I assumed must have been a meteorite. Yet, I could not understand why its trajectory seemed parallel to the surface of the earth. It was all over in less than ten seconds as I sat transfixed, in a complete state of shock, as it narrowly missed us on the port side. Looking over my left shoulder, I continued to follow its path. Stretching behind it, I saw a flaming tail, and when it reached the seven o'clock position, I saw it explode into spectacular burning fragments. It was over so quickly that I saw no reason to wake the two dozing pilots while I sat there in stunned contemplation of our narrow escape.

I am confident that if this burning object had hit us, it would have obliterated all evidence of our existence. History would have recorded us as another statistic of many other aircraft that have disappeared over large ocean expanses.

This event reminded me once more of our vulnerability. It is not only human error, mechanical failure, or inclement weather that stalk the skies and how our advanced technology is no guarantee of our safety.

Of course, this unusual phenomenon compounded my nagging concern, and Tony's lack of interest did nothing to dispel it.

Thankfully, an orange glow soon lit the eastern horizon. I was then relieved when the INS finally indicated we had arrived overhead the turning point, and the left-wing swooped down behind us to put the aircraft onto its new course. Once we headed towards the Namibian coast, we should have seen it at no more than one hundred nautical miles on the radar screen. However, we had to fly for over an hour before we even reached it, which meant we must have been nearly 400 nautical miles west of the correct turning point.

We had radioed ahead to Johannesburg when we left Abidjan and had given them our expected arrival time. Still, we left our unsuspecting welcoming committee hanging around at Mmabatho Airport until we eventually arrived, over an hour late.

Tony offered no plausible explanation for how we had flown so far off-course. However, once I recovered from the ordeal, I had a chance to study the chart and conclude that the only plausible explanation was that Tony had entered the wrong coordinates into the INS. Abidjan is a few degrees west of the Greenwich meridian. But, until then, the whole time, we had operated to the east of Greenwich. I am convinced that Tony must have made this error an automatic action after a tedious flight and done it after midnight. This wrong input into the INS explains everything as it would have then assumed our position was somewhere near Lagos. Unfortunately, our erstwhile instructor was in no mood to consider this possibility.

35 AFRICAN PRESIDENTS' WIVES ON CORPORATE FLIGHTS

Front entrance to Gulfstream GIII

While flying for Bop Air, one of my duties was to carry out corporate flights in our Citation II and Gulfstream III aircraft. The company policy was for the aircrew to have the plane ready and wait one hour before the passengers' scheduled arrival. However, as previously mentioned, I suspect their secretaries booked the flights an hour earlier than told. To compound this, the passengers themselves would most likely have told them to book an hour earlier than their planned departure time as well. Then, considering 'Africa Time', they were always late anyway. Their tardiness often resulted in the crew and aircraft waiting around for hours in the blazing sun. Moreover, we unnecessarily wasted fuel as we would have the APU running to keep the cabin at a comfortable temperature for when they finally arrived.

Being a born rebel who enjoyed the adrenaline rush of living on the edge, I would arrange for the crew to arrive only ten minutes before the scheduled departure time.

Over the six years, I operated out of Mmabatho, I only had one occasion when this practice nearly caught me out. We had landed at a small, isolated airfield between Brits and Ga-Rankuwa, where we had dropped off our passengers. A courtesy bus then collected us to spend the day at the Morula Sun Hotel. However, when a driver, who had just come on duty, came to take us back to the airfield, he failed to mention that he did not know where the airport was. We did not pay much attention to where he was going until we realised that he had not gone through the high-density suburb and must have missed the turnoff to the airfield. He was on the main road to Brits when I asked him where he was going. He replied, "I don't know." I suggested we stop and, only by recognising the small hill near the airfield, could we redirect the driver and manage to arrive only minutes before the passengers.

We often flew Bophuthatswana's President Mangope's wife and other family members around the country. We later carried the President and his wife in the Gulfstream III on trips to Israel, Munich, Dubai, and Rome.

The First Lady was a sour-faced, unfriendly woman, who looked more suited to living in a rural village but still assumed an air of superiority and disdain towards the aircrew. She frequently flew as the only passenger on short trips in the Citation. As this aircraft is too small to carry an in-flight cabin attendant, we would take two catering bags on board. We would fill one with miniature bottles of alcohol, packets of peanuts and snacks. We packed the other one with various cold drinks and ice blocks. Many of these short flights were to Wonderboom Airport, which only took thirty-five minutes. But before the President's wife disembarked, she would empty all remaining contents of the catering bags into her oversized handbag.

On one occasion, as she boarded the aircraft, she instructed us to delay our departure and asked her chauffeur to collect a vital package that she had left behind. When he returned in the official black Mercedes over an hour later, he handed it over to a waiting ground attendant, who could not resist a quick peek into the plastic Pick-'n-Pay packet, only to discover that we had been waiting for her hairbrush!

On their trips overseas, they were unashamed that it was to 'shop till you drop'. The two beautiful matching blue leather suitcases that accompanied the flight were so heavy that it required three people to lift them over two meters off the ground into the rear luggage compartment of the Gulfstream III. To get the suitcases into the aft hold, one person had to reach down from the upward-raised door with outstretched arms, while the other two worked together to lift them way above their heads to reach him.

On one trip to Dubai, they required us to reposition to Abu Dhabi to collect the Presidential couple and their entourage. A mountain of luggage and recent purchases from their lavish spending spree met us when we arrived there. The temperature hovered around 56 degrees Celsius while I hauled and dragged the baggage around the rear hold. The heat was so unbearable that the sweat pouring off my balding head drenched my sunglasses and made it almost impossible to see through the lenses. We were exhausted from the ordeal but had to tidy ourselves up, take up our seats in the cockpit, like ordinary 'lounge lizards', and fly the party on to Munich for a further week of shopping.

Although several immaculately dressed bodyguards went along on these overseas journeys, it never occurred to them to lend a helping hand, even though they put their bags with the rest of the luggage.

In Israel, they simply stood back to watch the pilots do all the work, despite our having been on the go the whole day, with a refuelling stop in between at the Nairobi Jomo Kenyatta Airport. They still expected us to hump the baggage onto trolleys, push them into the terminal building, and then to the front door of the five-star Tel Aviv beach hotel. Their standard sullen reply was, "It is not my job" if we ever asked for assistance. But I felt it was not the flight crew's job either.

When South Africa was heading for full enfranchisement and CODESA talks were underway to scrap the Apartheid system, we often carried leading figures from various political parties taking part in the negotiations. Our passengers would include the former military general turned politician, Constand Viljoen of the Freedom Front, members of the AWB, and, on several occasions, even Winnie Mandela herself. Most people are aware that she was the first wife of Nelson Mandela, who was soon to become the first black South African president. She was a big-boned woman with a well-cultivated charm and one of the most gracious passengers we ever carried onboard the Gulfstream. Furthermore, she was always courteous enough to introduce herself and shake hands with the cockpit crew whenever she came on board.

One trip we carried two cabin staff, Florence and Mapala, to care for our only celebrity. Winnie Mandela was the only passenger on the Gulfstream, with its luxurious, plush, grey leather reclining seats and polished rosewood tables. When Florence reported that the cabin was ready for takeoff, I noticed that she and Mapala were giggling about something. When I asked them what was so amusing, Florence explained that when she went to fold over the rosewood table in front of Winnie Mandela, they were amused when it exposed a fresh, wet, steaming lump of chewing gum. This still-moist lump had only just been stuck there by Winnie herself!

36 POODLE POPSICLE SWITCH

In a DC9 cockpit

In 1994, South Africa held its elections, and the former homelands of the Apartheid era were re-incorporated. Sun Air rose out of the ashes of Bop Air, the former national airline of Bophuthatswana. The company then relocated the airline from Mmabatho to Jan Smuts Airport, and with this move, purchased four McDonnell Douglas DC-9 aircraft.

I joined Bop Air in 1988 and progressed from the Bandeirante to the Brasilia aircraft. I also spent time on the Piper Chieftain and King Air 200B and did some corporate flying on the Cessna Citation II and Gulf Stream III aircraft. Towards the end of 1994, the company sent Ian McCleod, one of our new co-pilots, and me to St Louis, Missouri, to undergo DC-9 simulator training with Flight Safety International. The flight training system had a typically American flavour and style. The usual copious amount of coffee and doughnuts and accessible slot machines offered all sorts of chocolate bars, drinks, and other foodstuffs.

Part of the American training style incorporates a quirky way to help establish *the aide's memoir*. One example was how they emphasised the 'Poodle Popsicle' switch. Above the pilots ' heads, the designers had included it in a bank of switches on the overhead panel. They marked it in two positions: 'Fan' or 'Off.' When 'Fan' was selected, warm air was extracted from the cabin and directed to the forward cargo bay. The lecturers told us that whenever we transported live animals on the DC-9, they should travel in this forward hold and continually emphasised how crucial it was to select 'Fan'. The access to this hold, situated below the cabin floor, was by an outside hatch on the starboard side. They stressed dire consequences to the animal if the switch was left in the 'Off' position when flying at altitude, where the outside air temperature drops to well below freezing.

When most pilots start to fly, they are taught in meteorology lessons that the ambient (outside) air temperature decreases at approximately two degrees Celsius for every thousand feet gained in altitude. For general calculations, instructors taught us to assume that the temperature at sea level was 15 degrees Celsius. So, when an aircraft cruises at 30,000 feet, the outside air temperature will have decreased by sixty degrees under this assumed 'standard atmosphere'. In this case, the outside temperature would be down to about minus 45 degrees Celsius. Knowing this, one can picture frozen stiff animals at these altitudes if the pilots had neglected to heat the hold. Hence the pilots referred to this switch as the 'poodle popsicle switch', a popsicle being an iced confection in the USA.

After the lectures and simulator training at St Louis were complete, we returned to South Africa to finish our conversion onto the DC-9. Once checked out, I began to fly as a line captain between DF Malan in Cape Town and Jan Smuts in Johannesburg.

A few months later, I captained a midday flight to Cape Town with Ian McCleod, as my first officer. When the ground handler brought the load sheet to the cockpit, he pointed out that they had put two dogs in the forward hold. On hearing this, Ian remarked, "We mustn't forget the Poodle Popsicle Switch". I replied, "OK, will you switch it on please." The bright young ground staff, who often enjoyed sharing a joke with the cockpit crews, asked what the 'Poodle Popsicle Switch' was. As the passengers started boarding, I needed to concentrate on filling in the take-off speeds (V1 and V2) and power settings, so I asked Ian to explain this switch's function while I continued.

Once we had completed all the formalities, the cockpit door was closed, and we departed for Cape Town on a beautiful clear day. Since it was Ian's leg, he did the take-off after I had finished taxiing onto the runway. From that point on, he continued to fly while I did the radio calls, monitored the navigation log, and periodically recorded the engine parameters.

We had just crossed the Orange River, cruising at Flight Level 350 (approximately 35,000 feet above sea level) and had already passed Kimberly when Ian turned to me with panic written all over his face. Looking up at the 'Poodle Popsicle switch', he asked, "What position should this be in?" I answered, "On fan", to which he replied, "It's off then!" His alarmed expression of disbelief amused me as I replied, "Well, switch it on then." During our exchange, I noticed the look of horror on his face and beads of sweat breaking out on his top lip.

The flight time to Cape Town was usually about one hour and fifty minutes, including taxi time on the ground, and it had been a warm day in Johannesburg. By the time Ian saw the switch in the wrong position, we had been at cruising altitude for only half an hour, so I felt that the hold could not have dropped to the outside air temperature in such a short time. However, the look on Ian's face was more than I could take, and I could not resist winding him up a bit. If anyone carried the can, it would be me, as the captain of this flight. However, apart from putting the switch to the correct 'Fan' position, there was nothing more we could do at this stage.

There is a hatch on the cockpit floor to the left of the captain's seat, through which one could climb to get into the electrical bay. On the ground, with the gear down, one could even climb out through the nose wheel housing. I had previously heard a story about one captain in the USA who had snuck away in the 'e-bay' during a flight. When the chief cabin attendant reported that the cabin was clear for landing, the co-pilot asked her if she had seen the captain. She quickly carried out a frantic search but eventually informed the co-pilot that she could not find the captain anywhere. He told her he would have to land the aircraft by himself and said she could shut the cockpit door and take up her seat for the landing to complete the charade. When she had closed the door, the captain climbed back into the cockpit to fulfil his normal function until he had brought the aircraft to a stop in the parking bay. He quickly ducked through the hatch and climbed out through the wheel well onto the tarmac. When the chief cabin attendant opened the front door so passengers could disembark, she waited for the ground staff to push the mobile stairs into position. As she looked down, she nearly fainted from shock when she saw the captain standing next to the aircraft, smiling up at her.

As it was, I knew there was no access to the forward cargo hold from the electrical bay but could not resist saying to Ian, "I know, I will go down the hatch to give them mouth-to-mouth resuscitation." I then continued, "Or otherwise we could radio ahead to warn the ground staff to expect frozen animals on our arrival and to have a vet on hand with an oxygen bottle."

Ian had little sense of humour at this stage and could not see anything funny about my corny jokes. There were still fifty-five minutes of the flight left for him to endure, and by the time we were on the approach to the runway, he was in such a state of nerves, he could hardly concentrate on the task at hand and was shaking so much that he badly over-controlled the aircraft.

We finally came to a stop in the parking bay. When the baggage handlers started to arrive, and before we had even completed the shutdown checks, he opened his sliding window and, leaning out, shouted, "How are the dogs?" When he could get no sense out of the handlers, he barged down the stairs, pushing past the disembarking passengers, and dashed around to the cargo door to peer in.

On returning to the cockpit, he told me with noticeable relief that the large, highly pedigreed poodles were standing up happily in their cage when he got there. He was surprised to see there was also an untroubled cat on board! When he stretched his hand towards them, one of the dogs eagerly licked it.

Ironically, while all this was happening, I was surprised to see Mike Pfeiffer, one of our captains based in Cape Town, standing in front of the parking bay waiting for our arrival. Once the passengers had disembarked, he walked across towards the aircraft, and it turned out that he was there to meet the dogs, which belonged to his parents.

I shudder to think how Ian would have felt if he had realised his worst fears and the poodles had turned into popsicles! Especially if the first person to see them was one of our airline captains who had a personal interest in their safe arrival.

37 DC-9 FERRY VIA ICELAND TO THE USA

Sun Air DC9

I had been a captain on the DC-9 aircraft for less than a year when the Operations Manager called me in the first week of January 1995, to say he had booked me to fly British Airways to the UK the next evening.

Sun Air had recently purchased a DC-9 from Iberian Airlines, and a British company had fully restored it in the UK. The MD asked me to collect this aircraft from East Midlands and ferry it to South Africa.

The co-pilot, Dominic Murgatroyd, had been sent ahead of me and was already in the UK with an American aircraft technician called Johnny, a likeable character, about fifty years old, with a quiet sense of humour and undeniable experience on DC-9 airliners. He always stood out in a crowd, wearing a baseball cap or Stetson, cowboy boots, and a string tie. However, I never really discovered who Johnny's employers were, but I think that Sun Air used him in a liaison capacity to procure DC-9 aircraft.

Before I left Johannesburg, the Sun Air staff handed me a Business Class ticket, but when I arrived at the British Airways check-in counter, the check-in lady was Lucy Pocock, who had previously worked for Sun Air. She could easily have been a model with long legs and long thick eyelashes if she had not been so painfully shy.

After I handed her my ticket, she took some time to study her computer screen and then asked me to wait, and she disappeared on high heels over the luggage conveyor belt. When she returned, she told me to go to the British Airways passenger lounge and hand in my ticket at the reception desk and said they would call me later to issue me a boarding pass. Confused by this unusual procedure, I asked her if there was a problem, but she simply blushed and shook her head. When they finally called me to the reception desk, I discovered that Lucy had arranged an upgrade for me to the British Airways First Class cabin.

I had never flown first class on an overseas flight before and was more accustomed to the cramped discomfort of 'cattle class'. As I worked for an airline, Tish and I could take advantage of the staff rebate travelling facility. However, this meant travelling on standby tickets and suffering anxious hours, praying there would be *no shows*. We would then gratefully accept any available seat.

I endured a nagging pain above my right kneecap during my years flying the Alouette Helicopter. I often wondered if it was caused by continually pressing on the right yaw pedal and constant feedback from the helicopter's vibration. For years as an instructor, while the student flew the aircraft, I would sit rubbing this area above my kneecap to ease the pain. Whenever I flew overseas as a passenger, the only way to get some respite from the problem was to straighten my legs regularly. I also found it impossible to sleep if I sat with my legs bent. So, the luxury of having a first-class seat was a real bonus. Once aboard, I was like a child let loose in a candy store as I went through a basket full of movies I could watch at my leisure. It was such a treat to be able to select 'pause' and 'replay' at any stage. I also enjoyed the wide selection of refreshments, being able to stretch out and even lie sideways in lavish comfort.

At Heathrow, Johnny met me, and we travelled in heavy rain to a charming country hotel in the small village of Castle Donnington, close to the East Midlands Airport, where the other two crew members had already spent a few days.

Before I left South Africa, the ops manager told me that we would be ferrying the aircraft back along Africa's east coast. Still, by the time I arrived in the UK, there was already a hiccup over the mounted engines' purchase, and we had to wait there until Sun Air managed to resolve the problem.

The next day, I was taken to look at the DC-9 in the hangar and noticed it had a British registration. As the days dragged on, messages buzzed from one continent to another, trying to resolve the Pratt and Whitney engines issue. With time on our hands, we travelled around. I enjoyed the opportunity to visit a small nearby RAF museum where I looked over a Vulcan V-bomber parked on a grassy area amongst a few other retired aircraft. Apart from wandering around the quaint village with its rustic walls and visiting local pubs in the area, there was no more Dominic, or I could do but wait for further instructions.

In the meantime, as the aircraft did not have a GPS, Johnny bought a handheld Garmin GPS instrument, which came with a short length of cable to plug into its aerial socket. This extension, which could be attached to a suitable part of the cockpit near the windscreen, would hopefully enable us to pick up signals from relevant geostationary satellites. So, while we continued to wait, I took full advantage of the delay to familiarise myself with its instruction handbook.

After a week's delay, Johnny told me that they had changed the whole plan. He said we would now have to ferry the aircraft to Springfield, Illinois, routing via Iceland, Greenland, and Canada. So, Johnny and Dominic went back to the British Airways offices at Heathrow to collect the necessary aeronautical charts and let-down plates for the route. Although we were more familiar with the Jeppesen system of Navigational logs and charts, they could only get the Aerad version of aeronautical plates.

Once they had established our departure date, Johnny arranged with one of the catering companies at the British Midlands Airport to provide us with food and drinks for the flight. In the meantime, I worked out the flight plan and fed 'waypoints' along the route into the little GPS. I also drew each track onto the aeronautical charts and filled the whole route's details onto our company navigation logs. Once completed, I felt happy that this would help us keep track of our headings, times, and fuel consumption. We then submitted a flight plan for the following day, with an expected departure time of 0300 Zulu.

When I finally arrived to check out of the hotel in the early morning, I felt I had not slept a wink. After spending hours poring over the procedures and frequencies for the departure and the anticipated let-downs at Reykjavik and Soderstrom, I was so overstimulated and waiting for the early wake-up call, that it had been difficult to get a good night's sleep. The hotel's receptionist handed us a fax from Johnny's head office in Springfield, Illinois, to add to our already stressful situation. It informed us that they could not get landing clearance in Canada, and we would now have to route via Goose Bay in Newfoundland.

This message threw us a curveball. But thankfully, just before departure, we managed to send a fax to the head office, asking them to forward approach plates for Goose Bay to collect at Reykjavik's landing office so that we could alter our flight plan there.

Once strapped in and ready to start, it felt strange to operate out of unfamiliar surroundings and still call out our checks and responses from our regular company 'checklist'. It was the first time I had a chance to sit in this aircraft, and it felt more like being in a DC-9 simulator, where the instructor had the facility to change our location to anywhere in the world. It was surreal and difficult to absorb that we were about to head off towards the North Pole, passing over Scotland and the black, freezing North Atlantic Ocean. I prayed that the GPS would behave itself, as once we passed the point of 'no return', there would be limited loiter time over Iceland and insufficient fuel to divert to an alternative destination.

I decided to fly the first leg, so Dominic made the necessary radio calls. He confirmed our track by cross-referencing with NDB and VOR beacons along the route until we left the security of Great Britain's aviation facilities. Far too soon, we were out over the inky black North Sea, leaving the lights of Scotland behind us.

I monitored the GPS with the autopilot engaged as there was nothing more to go on. Fortunately, the instrument behaved well and acquired multiple satellites along the route.

After what seemed an interminable period, heading into total blackness, Scottish Area Control instructed us to call Reykjavik 'within range'. Still, out of radio range, we sat in silence in the dimly lit cockpit, making minor heading adjustments on the course selector to keep on the 'great circle' track across the ocean.

At length, we established contact with Reykjavik, who advised us of reduced visibility in blizzard conditions. They told us that the main runway, with the only ILS approach facility, was not available to my absolute horror. The other runway only had a VOR to assist us with the let-down. Before 'top of descent', I told Dominic to make this a 'monitored approach.' He would have to take over as the 'flying pilot' from the 'top of descent' and fly on instruments until I had the runway lights visual, at which stage I would take over for the landing.

We studied the chart and safety heights before going through the let-down briefing and over-shoot procedure. The let-down plate indicated that it would be a 'stepped down approach', in which I had to control and direct Dominic on what headings to steer and when to descend to the next level before commencing our final approach onto the runway. I closely monitored the distance from the field on the DME readings and bearings on our VOR instrument to achieve this.

We descended to the first altitude and reduced our speed in preparation to select the flap and gear. I advised Dominic when to continue down to the next level at each checkpoint. We configured the aircraft for the landing at the appropriate distance, but he was slow to establish the correct descent rate to reach the four nautical mile checkpoint's required height at nine nautical miles. To avoid a go-around and re-position for another attempt, I called for a 1,500 feet/minute rate of descent until I instructed him to resume 600 feet/minute. We followed the pictorial representation on the small let-down plate. We flew in the cloud, with total darkness outside and no visual picture of the terrain below. Although I did not stop to dwell on the situation, I must confess that I felt uncomfortable and prayed I could keep abreast of our progress and re-establish the aircraft on the correct profile. Thankfully, we achieved the correct height by four nautical miles, and I breathed a sigh of relief, feeling exonerated that we had managed to salvage a precarious situation.

Continuing from four miles at the proper rate of descent, Dominic called, "One hundred feet above," and then, "Minimum! Decide". At that very instant, although the driving blizzard of snow was almost blindingly bright as it reflected the landing lights at us, I picked up the lead-in lights, glowing with blazing intensity. It was the first time I had ever seen 'running rabbits', which are pulsing strobe lights that look like a single white ball of light moving toward the runway threshold. It was such a welcome sight as I responded, "Continue". Although the visibility never increased to further than 500 meters, in no time I picked up the runway threshold lights and a few runway sidelights and took over for the landing. Landing in this blizzard felt like landing in thick fog, where you think you will catch up with an area of zero visibility, but it continues to keep just ahead of the aircraft. With the outside air temperature well below freezing, it was impossible to do a hard landing or 'thump it in' on this surface, as it was smooth as an ice rink.

Once I reduced the aircraft down to taxi speed, I noticed the snow piled up to 30 feet on either side of the landing area, and as we cleared the runway and Dominic carried out the 'after landing checks', I turned to look at Johnny, who had been sitting between us on the jump seat. The let-down had been extremely tense, and I felt my heart pounding in my throat, but we acknowledged our shared sense of relief without a word.

I asked Dominic to go and pay the landing fee and change our enroute Flight Plan while Johnny went to collect the faxed approach plates for Goose Bay. In the meantime, I attended the refuelling and carried out the aircraft's external 'walk round' pre-flight check. It had stopped snowing by then, but it was still a dark, unknown world, away from the lights of the terminal. The temperature on the ground was minus 30 degrees. The sub-zero temperature had the strangest effect on the inside membrane of my nostrils, which crinkled like crepe paper with each breath I took. I was glad to get back to the warm glow of the cockpit, as the auxiliary power unit we had left running kept pumping in hot air, with its jet blast roaring out the back of the fuselage. The unbelievably low temperature and the dark world around us were such strange and unfamiliar environments, we may just as well have been on another planet.

With all the critical approach and let-down plates for Goose Bay in hand, we felt more relaxed as we set off on the next leg to Greenland. It also gave us a more secure feeling, knowing that we could trust the little handheld GPS, which had done such a fantastic job so far.

At Soderstrom, the ILS let-down was between the fjord's steep sides with the runway nestled at the base. Studying the 'let down' plate, I noted that the overshoot was a 'Procedure Turn,' which generally requires the pilot to turn 30 degrees off the line of approach and hold this heading for one minute, before turning back to overfly the facility and exit on the reciprocal to the approach path. The surrounding high ground of the fjord made this the only possible escape route. Fortunately, this time the weather was much better, and we could see the runway from 800 feet, so I let Dominic take over for the landing.

Once again, I did the walk around, and by then, the outside light improved just enough for me to see the backdrop of steeply rising ground just behind the terminal buildings. I was horrified to see just how close it reached above us. It was the middle of winter, and I was amazed when someone told me that it was the most daylight, they would get that day. To me, it looked more like a moonlit night, with less than a half-moon.

From Soderstrum, we flew on to Goose Bay in Newfoundland and arrived on a crisp, clear day, with fresh white snow piled up high on the sides of the runway and taxiways. After a quick refuel, we flew to Buffalo Airport, near Niagara Falls, where American Aircraft Traffic Control Officers welcomed us in their familiar, efficient, and professional style.

Later that day, which happened to be a Saturday, we

could see Buffalo with clear visibility of over ten miles. On our descent, we could see the airfield, situated on a large stretch of flat ground, and could make out other aircraft looking like toys, flying in slow motion, around the vicinity.

We had enjoyed our packed meals along the way, with Johnny acting as the cabin attendant, heating various tin foil containers in the galley oven. They were still not finished, and the caterers had also spoilt us with oranges and other fruit, which I was saving for later. The customs officials who met us on the ground cheerfully asked us if we had any fruit on board. I presumed they were scavenging in my innocence, but instead of wanting something to eat, they tipped the whole lot into a plastic bag to be burnt, as they happily informed us it was illegal to bring fresh fruit into the United States. If I had known this, I would have finished it off before our arrival or hidden it away in some storage place. As it was, it was painful to see it all go to waste.

By the time we arrived in Buffalo, we had already been on the go for more than twelve hours and just wanted to get this trip over with, but they asked us to wait for the chief customs officer before they could clear us to continue to Springfield. If this had been in some part of Africa, I would not have been surprised, but since this was a 'First World' country, I could not understand the reason for the delay. The customs officials were deeply sorry for the delay and invited me to sit in their office while waiting. Dominic and Johnny stayed on board the aircraft and managed to catch up on some desperately needed sleep while I hung around the office, unwashed and unshaven, with my eyes raw and gritty and my brain numb. Barely staying awake or concentrating enough to make polite conversation, I sat at a computer playing card games for the next few hours.

When we were finally on our way again, it was after sunset, and, for the first time, the GPS refused to pick up any satellites. The lack of GPS coverage meant searching for navigational beacons on our Aerad aeronautical chart. In retrospect, the satellites' concentration, which was possibly directly above us, may have been screened by the cockpit overhead bulkhead. It would have been far more accessible to key in the VOR beacon's identification letters and press 'Go to' on the GPS.

Most of the time we had been flying, it had been dark. Now it was dark again. I became so confused that every time I contacted a new radar air traffic controller in my state of fatigue, I kept greeting them with "Good morning". This greeting might have seemed strange to them but not too surprising, considering my foreign accent.

Dominic felt back in his element as the 'flying pilot' with a plethora of beacons all around us. He began to ask for cross-references to every available beacon around us. We had just flown for hours over the Atlantic, where the only navigation source had been the handheld GPS. Now I was tired and in no mood to indulge Dominic's every whim and snapped at him, "Listen, Dominic, just give me a break".

After that, he kept quiet, which allowed me to attend to the task at hand of complying with air traffic instructions.

By the time we reached Springfield, we had discovered from the ATIS (voice weather recording) that there was total cloud cover with low stratus down to 200 feet above the field. This low cloud base entailed another monitored ILS approach down to 'minimum'. If we had not been able to land off this approach, I am not sure what other options were available to us, as we had no let-down plates to divert to anywhere else in the USA. Adrenalin kicked in to revive me for this final burst of concentration, and I let out a breath of gratitude when we broke out of the cloud, right at 'Decision Height', and saw crystal clear runway lights ahead.

Springfield was Johnny's home base, and we had a welcoming party of well-dressed gentlemen and their wives who whisked us off to our hotel in a brisk, matter-of-fact manner. Utterly exhausted, I was relieved to have made it through the ordeal with no mishaps along the way, as I was fully aware of how much we had stretched the envelope and how easily things could have 'gone incredibly wrong'.

Dominic asked if I would like to "hit the town" en route to the hotel, but I declined, with the combination of my more advanced years and feeling jaded. After a quick room service snack and shower, I happily crawled straight into bed. It was so soft and inviting that I was asleep as soon as my head touched the pillow.

38 DAVE KUHN – A MEMORABLE CO-PILOT

Dave arrived on the scene at Sun Air, directly from the South African Air Force and joined as a First Officer. Sun Air had been operating the DC9s out of Johannesburg, Cape Town, Pilanesberg, and Durban for a few months after Bop Air had been renamed and moved its base to Jan Smuts. The new government body reincorporated Bophuthatswana into South Africa after April 1994.

Dave had a square-faced Germanic good look with a tendency towards becoming overweight. He had thick straight, light brown hair, with a neat wave held in place with gel, more reminiscent of the 1950s style. This first officer was also annoyingly intelligent, smooth, and charming with a wicked, mocking spirit.

I battled not to show how quickly Dave could get under my skin. By his obsequious charm and flattery, he could suck me into confiding my innermost thoughts and history. He would later use this knowledge as a weapon during a moment of weakness.

What probably got to me more than anything else was his boyish charm with the cabin staff, with whom he took every opportunity to try and embarrass me. His annoying sense of humour could quickly descend into a childish game of 'one-upmanship'. Most of the time, I fought to hide my irritation, and I know it would have been better if I had been more mature and aloof, but this was not in my nature, and our relationship in the cockpit was more one of love-hate. He also sensed that I took too much pride in my landings, which I found annoyingly inconsistent on the DC9, with its tapered wings providing no cushioning effect as it came close to the ground. Having always taken pride in my flying ability, I was starting to feel a drop in my previously high standards and often became humiliated by Dave's derisive roar of laughter whenever I 'fluffed it'.

He took great pride in his appearance and dressed immaculately. He claimed he would never go bald, in contrast to my diminishing hairline. However, he was exceedingly vain, especially about the fact that he had inherited his dad's thick hair and took great care with it, ensuring that he slicked back each greasy strand. Now and then, I could not resist the temptation to ruffle his thick mop of hair and enjoyed watching him scurry off to the aircraft toilet, with his readily available comb, in case any of the cabin staff caught sight of him in a messy state.

He also savoured every opportunity to make the passenger announcements, which he did with sickening professionalism. He would make it especially enjoyable by pointing out every notable feature along the route and passing on titbits of technical data about the aircraft's performance, including details like how many litres we would burn on the flight and the outside air temperature. His other significant weakness was his burning ambition to become wealthy and how he envied anyone with affluence, coveting their cars and other status symbols.

I was discerning this as his Achilles heel, which allowed me the chance to wreak my revenge. To make the passenger announcements on one of the DC9s, pilots had to do it over a handheld phone after pressing an overhead blue light switch before he could start his spiel. On one flight, when he was the flying pilot from Johannesburg to Durban, as we came alongside the sprawling Vaal Dam, with its many twisting and spindly inlets, he activated the switch and started his waffle. Soon after he began to talk, I surreptitiously disengaged the phone by depressing the light button. When he finished his lengthy preamble with evident smug satisfaction, I asked him if he enjoyed talking to himself. His eyes darted to the top console in disbelief when he saw the light was off! Completely undeterred, he then went through his entire speech again while covering the switch with his left hand!

I feel obliged to confess with hesitation and embarrassment that he caught me out on two separate occasions. The first time was during a flight from Johannesburg to Cape Town, just as I emerged from the toilet near the front galley. I found it comforting and therapeutic to drink copious amounts of tea on each flight, and the cabin staff happily indulged me. Unfortunately, this also meant quite a few trips to the toilet, but I enjoyed the chance to stretch my legs anyway. I had become accustomed to Dave pulling out the circuit breaker for the toilet light and then rocking the aircraft in the hope that on my return, I would have messed up my spit and polished shoes. On this occasion, as I emerged, I was intercepted by Kim, who was the chief cabin attendant and one of my favourites, whom I thought would have shown more loyalty. She asked me if I could assist one of the cabin attendants halfway up the aisle between the passenger seats in the economy section who was having difficulty with her trolley. I always felt a bit conspicuous if I ever had to enter the passenger section of the aircraft and started down the aisle. As I approached the other young attendant at her trolley and asked what the problem was, the sad look on her face alerted me to the fact that they had duped me. Sure enough, Dave's dulcet tones oozed out over the loudspeakers, "Ladies and gentlemen, I would like to introduce Captain George Wrigley to you. Despite appearances, he is not the oldest captain on the fleet, and he is now going to do an impersonation of Elvis Presley for you." He had well and truly caught me! As the passengers responded with obvious delight at my embarrassment, I continued to the back, shaking hands along the way, until the last seat, where a giant African American asked me in his Southern drawl, "Well, Man, what are you going to sing?".

When I finally managed to get back to my seat in the cockpit, with Dave crying with laughter at having caught me out, I went onto the cabin address system and said, "Ladies and gentlemen, I must apologise for my co-pilot's childish sense of humour but here is:" and started singing,

> "Well it's a one for the money, two for the show,
> Better get a ready now go cat go
> But don't you step on my blue suede shoes."

When we had parked on the hard standing in Cape Town and switched off the seat belt signs for disembarkation, every passenger put their head in the cockpit door to say goodbye, and some even said it was the best flight they had ever travelled.

I feel humiliated to reveal that having been caught out once before by Dave and the cabin crew, I fell for the same ruse a second time when the chief cabin attendant asked me to help with an unruly passenger.

As I entered the cabin's economy class section, I heard a muffled crackle coming from the loudspeakers. It was Dave's sonorous voice crooning, "I'm too sexy for my shirt…." but fortunately, I managed to turn on my heels before anyone else realised what was happening.

A little while later, on a trip back from Durban to Johannesburg, while passing Pietermaritzburg, I casually mentioned to Dave, "On our right-hand side is Pietermaritzburg and just up that tar road going north is Wartburg. If you look a little beyond, you can make out the estate where I grew up." My piece of information had an immediate and extraordinary effect on Dave, as he asked with wide-eyed wonder and previously unseen respect, "You never told me your family has an estate. Who lives there now?" "Oh, just my mother and aunt, but we have a manager looking after it." I had visited the area when my sister's husband worked in a sawmill there so I could make up the story as I went along.

To his protests that he thought I came from Zimbabwe, I responded that I only went there when I joined the Air Force. He was also impressed when I informed him that I went to Michael House, an elite private nearby school. When he asked me, who would inherit the property, I said probably me. However, I was sure my sister would also get a lot of the family fortune. He then asked how much the estate was worth; I replied in an offhand manner, "Probably about five hundred million Rand."

The first time Dave visited our home in the up-market village of Irene, near Pretoria, he immediately recognised the quality of a few pieces of antique furniture that Tish had inherited when her aunt had died in the UK. These pieces complemented the Rhodesian teak parquet flooring, which we had restored with a gloss varnish. I had also painstakingly removed layers of paint that previous owners had applied to the beautiful golden-orange Oregon pine door frames and skirting boards over the last 60 years. It was a gracious home built by a cousin of General Jan Smuts. It had a magnificent English country garden, abounding in roses, wisteria, and banksias, and full of indigenous and exotic trees. This home had made an impression on Dave because after the flight, as we signed off in the operations room, he turned to me with, "You never had me fooled. I always knew you were wealthy!"

Al Bruce, who was then the chief training captain, enjoyed the charade and, on a subsequent flight, left a message for me attached to the flight logs, which the co-pilots usually collected. It read, "George, your aunt rang from the estate. She says the Rolls is due for a service." This impressed Dave immensely, as status symbols were so important to him. I enjoyed my improved standing, which emanated from this fabricated wealth. Amazingly, this continued for another few weeks before someone finally let the cat out of the bag. Of course, Dave was furious at have been so thoroughly hoodwinked.

My attitude towards Dave softened one day on a return flight from Cape Town to Johannesburg when he told me his father was coming on board and had asked permission to join us on the jump seat. I knew he was a retired SAA captain and welcomed the opportunity to get to know him. I always enjoyed meeting passengers and readily agreed.

During the flight, I found Vince Kuhn polite and humble and saw another side of Dave. As we approached Parys, we needed to manoeuvre around a thunderstorm line with colossal cumulonimbus clouds towering above us. Dave suggested we go east of a particularly vicious-looking build-up, with an overhanging anvil. However, I knew from experience that this was often an area of extreme turbulence with possible hail. Experts had always advised to avoid flying anywhere near it. Instead, I recommended we go to the left as I knew their movement direction was predominantly west to east. It then gratified me when his father concurred with my suggestion. But what touched and impressed me most was when his dad said goodbye to Dave, who had to remain in his co-pilot's seat until the passengers had all disembarked. Vince simply leant forward and kissed Dave on the lips. This sweet, unself-conscious gesture moved me.

After a few years of this love-hate relationship, South African Airways offered Dave a job. So, we had a farewell party for him at the Company's Jet Park office pub, where I had an opportunity to make a farewell speech.

I was then touched when his wife commented that she hoped that other captains in SAA would be as friendly and kind to Dave as I had been!

39 GIRL PILOT – MAXIE

Maxie was the only woman pilot I had ever flown with, and it was a unique experience in many ways. It certainly brought a new atmosphere into the cockpit and the cabin staff and passengers always seemed to enjoy her presence.

She was quite pretty, but her pride was in her most outstanding feature; her long dark red hair, which fell below her waist. Although, to conform to the same standards set for the women cabin staff, she had to keep it tied up and off her collar. She had a sweet soft voice with a faint Afrikaans accent, which sounded almost Gaelic, and the passengers loved to hear her announcements from the cockpit. These messages followed the style adopted by other Afrikaans co-pilots, who would give the whole spiel in English and then repeat it in Afrikaans. I always felt it was a waste of time as everybody understood English anyway. I once asked her why she did not do it in all the fourteen official languages of the new democratic South Africa, but she could not see the irony in that.

Maxie started with Sun Air as an operation assistant and managed to win everyone over, including the Operations Manager and the Managing Director, with her gentle and coy manner. When they learnt that she had a private pilot's license and was busy working towards a commercial pilot's license, the MD decided to let her train as a co-pilot on the Citation II aircraft, to build up her flying hours. So, whenever a job came up, she flew with Loutjie Naude, one of the company's training captains.

From the Citation, she progressed to the DC-9 and was sent for her simulator training to Flight Safety International in St. Louis, Missouri, and then continued as an online first officer with a training captain.

Coming from a military background, it took some getting used to pilots who had not gone through the same strict military protocol during their training. In Maxie's case, she further developed her relationship with the cabin staff by using her feminine charm and guile. Her manner was in keeping with many of her other Afrikaans sisters and young wives, who had to keep their much more muscular and virile men in check. What I found incredibly challenging to accept was the independent way she ignored any instructions offered by the captains, like, "Don't try and make the first turnoff" after every landing. Still, she would continue to hammer the brakes to make the first available exit. Otherwise, she would use up most of the runway to put down a *'greaser'*.

One of the training captains was a French-speaking Belgian from the old Belgian Congo, a caffeine and nicotine addict. He suffered severely from nicotine withdrawal symptoms on any long trip; he found it exceptionally hard to cope with Cape Town and Johannesburg's flights, as he had to comply with the Sun Air 'no smoking' policy. The withdrawal symptoms that set in made him unusually irritable by the end of the day.

Maxie was his co-pilot on one occasion, and she ended up floating halfway down the runway, determined to put down a smooth landing. He had told her to put the aircraft down, and after being ignored twice, he could take it no longer and shouted at her in his gruff French accent, "Put the *f...ing* aircraft on the ground!"

Whenever Maxie crewed with me, I had to endure the cabin staff girls congregating in the cockpit, while waiting for the passengers to come on board, and would take turns to brush her beautiful tresses of red hair.

However, I needed to retract every moment of previous frustration I had experienced when flying with her, when she put down a landing on runway 21 Left at Johannesburg International Airport, after a cloud burst. The torrential rain during the thunderstorm had left the runway covered with water. This runway rises to a high point and then slips quickly down towards the south of the field, and once again, she had held off far too long so that the aircraft only touched down the other side of the highest point. When pilots turned the plane off at the end of this runway, they could see the East Rand Mall shopping complex on the other side of the freeway. When they constructed this runway, they built a ramp to increase its length and make it as level as possible. However, this left the last hundred meters as a grassed extension, before it fell sharply into a vlei area just short of Benoni and Johannesburg's motorway.

She applied the reverse thrust on the two levers in front of the thrust levers, and as the aircraft decelerated to 100 knots, I took over control as the steering wheel was on the captain's left-hand console. At 80 knots, I disengaged reverse thrust in keeping with the company's standard operating procedures, as this avoids ingesting water and other loose objects into the engine intakes. I also started to apply the brakes on top of the rudder pedals.

The hydraulic brakes have an anti-skid mechanism to prevent the wheels from locking under slippery conditions, which could result in a burst tyre, especially in the case of water patches on the runway. As I tried to apply brakes, it was immediately apparent that the anti-skid mechanism was in constant operation, as I could not get any further reduction in aircraft speed. I realised in horror that we were not going to stop the aircraft before the end of the runway. As everything went into slow motion, my mind was rapidly going through thoughts of the embarrassment if we ran off onto the soggy grass and, at best, have the wheels sinking to the hubs, but even worse, if it culminated in the aircraft tumbling down the steep embankment. I also thought of the 'occurrence or accident' reports I would have to write up, as the captain is always ultimately responsible for the aircraft, crew, and passengers' safety. I was already preparing myself for the passengers' apology, shutting down the engines and calling the Company operations room to arrange for a bus to collect everyone from this far end of the airfield and transport them back to the terminal.

While we slid, I asked Maxie to assist me with the brakes. We both stood on the pedals but to no avail until Maxi, using her common sense, feminine practicality, and lateral thinking, suggested we try to reverse thrust again.

Until that moment, I had been far too disciplined and frozen to consider this as an option. But, as soon as I pulled back on the reverse thrust levers, I felt the immediate deceleration. Full engine thrust was now deflected forwards by the two sets of clamshells, closing together behind the engine jet pipes. This reversed thrust sent water on the runway surface from beneath the engines, in a forward blast of spray.

With heart-stopping relief, I managed to stop the aircraft just in time to taxi off at the end of the runway at taxi speed and call for the after-landing checks.

If possible, I would have immediately hugged Maxie in gratitude, but I made sure that I acknowledged that she had saved the day, with her quick thinking and calm practicality.

I cannot say that Maxie never frustrated me again or that I ultimately overcame the feeling that she lacked an innate sense of discipline and respect towards the aircraft captain. But I will be forever in her debt for keeping my clean accident-free record intact over my thirteen years with Sun Air.

40 WINDSHIELD ANTI-ICE

During our line-up checks, I can no longer recall how and why we missed selecting the windshield anti-ice. I think it may have been due to the excessive outside air temperature that day and trying to keep it a bit cooler in the cockpit during a long delay, while waiting for our turn to take off. But it was still a matter of 'finger trouble', as we used to say in the Air Force.

It all took place on a day that I was flying as the captain of the McDonnell MD80 aircraft from Johannesburg to Cape Town, with a charming first officer, Johan van Niekerk, an ex-Mirage III pilot from the South African Air Force. We had flown as a crew several times before this incident, and I had always found him a person of natural humility with a professional approach to his flying. He was also a good-looking young man, who did not go unnoticed by our young cabin attendants.

On this day, the trip went off routinely but, typical of the South African climate, when there is clear weather on the high veldt one will likely experience inclement weather at the Cape or vice versa. We had sailed along, happily unaware of the impending drama and enjoyed the attention of a bevy of cabin crew, who visited the cockpit to bring us lunch and tea or just for a chat. The skies were clear, and the autopilot took care of the flying and navigation. As the non-flying pilot, Johan made the routine radio calls and filled in the flight log at checkpoints along the route.

We checked the ATIS recording, which gave us an update on the current weather conditions at Cape Town and then established radio communications with Cape Town Area Control at their boundary. The reported weather was low stratus with fog over the Cape Flats, and the visibility was reduced to five hundred metres. These weather conditions necessitated a monitored approach and landing. So, we went through the briefing, and Johan took over as the flying pilot.

The first officer flew the let-down on instruments while the captain monitored the progress on monitored approaches. In the last stages of the ILS approach, at one hundred feet above the 'decision height' of the altimeter setting, the first officer would make the call, "One hundred above," to which the captain would respond "Check", after confirming with the 'bug' on his altimeter. The next call made by the first officer would be "Minimum – decide", at which stage the captain would need enough outside visual reference to continue with the approach. Whether he felt the visibility was good enough to land or not, the captain would either respond with, "Visual – continue" or "Go around!" In either case, the first officer continued to fly on instruments. After making the call, "Visual, continue," the captain could still, at any stage, call, "go-around". However, once he was fully satisfied it was safe to land off the approach with a good view of the lead-in lights and runway threshold, he would announce, "I have control". At times, in fog, the captain could only see a few hundred metres down the runway at any stage. He would have to keep the aircraft on the centre line, with the nose-wheel steering, while applying reverse thrust and brakes, and decelerate the plane down to taxi speed before he could turn off the runway when the relevant exit came into view. Once off the runway, the first officer would inform the tower controller, "Clear off the runway." Under these conditions, the tower controller, who might not be in visual contact with the aircraft, would instruct the plane to call ground control for further taxi instructions.

We completed our briefing, and just before setting

up the descent, the chief cabin attendant informed us of a passenger's request to sit in the jump seat so he could watch the approach and landing. I permitted her to bring him forward, as I always enjoyed meeting and relating to the passengers and considered it good public relations.

We passed overhead Sutherland NDB, and shortly after that, we asked for descent clearance. The first clue of any problem came at 14,000 feet as we passed overhead Paarl. The aircraft popped through a thin layer of cloud. As that happened, a thick layer of ice slapped onto the windscreen and blocked all forward visibility. In unified shock, we instantly looked up at the bank of switches above our heads, and to our horror and disbelief, we discovered that the windshield anti-ice switch was in the 'off' position. Although there were still clear blue skies outside the aircraft, we now had to fly on instruments.

Of course, we quickly selected the switch to the correct position. But since we had spent nearly an hour at sub-zero temperatures, I knew it would take some time to melt the build-up of ice, which now totally blocked all forward visibility. It would have been impossible to carry out an approach and landing under the present condition, so I requested clearance to head towards Robben Island to sort out a "Small technical problem". Cape Town ATC approach gave consent and asked if we needed any emergency crews to stand by on the runway. I declined the offer as this would entail submitting a report to the Department of Civil Aviation. Apart from the mountain of report forms I would have to fill out, I felt we should avoid any further embarrassment.

After ten minutes, the windscreen cleared enough for us to feel confident that when we were on the final approach in the warm air at a lower altitude, I would see the runway when we broke through the low cloud base. We reported that we had sorted out the problem and that we were ready for the approach and landing. The approach controller gave us headings to steer to establish us on the straight-in approach for Runway 01. He directed us to the east of the airfield, towards False Bay, before turning us onto finals for the ILS let-down. He then cleared us to Tower Control with instructions to report "Established at the outer marker" and "Visual".

I was not sure how much the unwitting passenger who had joined us in the cockpit, was aware of our problem. However, we were so busy handling the situation that we ignored him. We certainly had no time or intention to bring him into our confidence, and although he wore the spare headset, we spoke in either veiled terms or technical jargon, which he was unable to follow or understand. However, I can only assume that he was nervous, as it was the first time I was aware of an acrid odour, which I guess was the smell of adrenalin from his sweat-soaked underarms, and I also saw sweat pouring down the sides of his cheeks. Regardless, it was not a pleasant odour.

Although the windshield panels in front of each pilot had cleared, we did not notice that ice still blocked the windscreen's centre panel. This ice made it impossible for the passenger to see anything out in front of him.

We were relieved to have extricated ourselves from an awkward situation without further complications. We remained unaware of the restricted view for the passenger in the jump seat.

The runway came into view in the thick swirling fog at the decision altitude, and I conducted a routine landing, considering the circumstances. We then taxied off the runway and finally caught sight of the marshaller, who waved his wands to guide us into our allotted parking bay.

Once the shutdown checks were complete, we could finally breathe a sigh of relief and turn our attention towards our passenger for the first time since he had joined us in the cockpit. We only then noticed that ice had remained on his part of the windscreen, and our passenger was still unable to see anything out of the front portion of the aircraft windscreen.

We had to hold back our burst of hilarity and relieved laughter until after he had vacated the cockpit. Especially after he thanked us and added, in utter amazement, "I really don't know how you guys do this!"

<u>Footnote on Johan van Niekerk</u>: Johan later became interested in flying the Pitts Special and was busy training to become part of the Rothman's aerobatic team. Sadly, he and his passenger from the UK had a fatal accident while doing some low-level aerobatics.

Before this accident, Air Emirates had accepted Johan's application to join them. He and his family were busy gearing themselves up for an exciting new life in the Middle East.

42 THE LAST SUN AIR FLIGHT

MD 80

The date was Friday 13 August 1999. I joined the day's last flight from Johannesburg to Cape Town as the Captain of the MD 80 aircraft. My co-pilot and I were Johannesburg crew members and were due for a night stop once we arrived in Cape Town. However, all the cabin crew were based in Cape Town and were flying back home.

After getting on board, a message was sent to us by the operations room. It told us that South African Airways had bought out Sun Air and would liquidate it the next day. Of course, this news came out of the blue and put us all in a state of shock.

At the end of the Apartheid era, South Africa reincorporated Bophuthatswana into its fold. We had been worried about the future of Sun Air four years before, when Comair tried to block us from starting a service from Jan Smuts to Cape Town and Durban in 1994. But now, our airline seemed a well-entrenched and highly respected airline, and nothing had given any hint of this possibility.

We later learnt that Sun Air's MD Johan Borstlap had arranged for someone to take videos of each of the airline`s operating divisions for his record, two weeks before. His request seemed rather strange at the time, but apparently negotiations were already well underway for this whole exercise to take place. Safair, who was leasing the MD80 aircraft to Sun Air, was also a party to these negotiations.

The liquidation message included a direct threat that we had to continue flying until the last flight the next day. If anyone refused to cooperate, it would affect their retrenchment packages. Even so, the senior cabin attendant on the flight approached me to inform me that the cabin staff had discussed the situation and decided not to work on their return flight to Cape Town and would travel as ordinary passengers. In reply to this, I said, "OK, that's a good idea. Since my base is in Johannesburg, I won't fly either! Bye, girls. See you sometime!" Well, the response was immediate. They begged me to continue with the flight, saying they would carry on with their duties as well.

When we arrived at Cape Town that evening, the cabin attendants removed the entire bar stock from each landing aircraft as it shut down and brought it to the operations room, where an impromptu party developed. It was surreal, and although I have never been to a wake, it gave me some idea of why people hold them.

The next day I had four more flights between Cape Town and Johannesburg and did the last Sun Air flight that left Cape Town that evening.

The flights continued with a more celebratory than a grieving atmosphere the whole day. Over the previous few years, Sun Air had expanded to over three hundred personnel, who had grown into a close-knit family. I suppose knowing that we would not be working together again gave us the freedom to express the *'Esprit de Corp'* and affection that had developed between us.

When the check-in staff handed us the passenger manifest for the last flight, I noticed only one lady in business class and eight other passengers. So, I asked the senior cabin attendant to ask this lady if she would mind if we moved all the passengers to join her in the business class area. When I knew that she was happy with the arrangement, I announced to the passengers that this would be Sun Air's last flight and we wanted them to feel like VIPs and move into business class. I asked that if anyone had objections, please inform one of the cabin staff. After takeoff, I alerted them that I would turn back for a fly past over the airfield as a final farewell.

As it turned out, everyone was quite happy with the whole arrangement, so I asked the tower controller for clearance to carry out a fly past just outside the security lights of the apron area. Like so many others, the controller, whose voice had become recognisable, was happy to accommodate this request and arranged with the approach control to vector all incoming traffic to keep them clear of my intended departure.

After taking off at twilight in a southerly direction, I climbed 1,000 feet above the ground. Once cleared by the tower control, I carried out a teardrop procedure to return parallel to the runway and flew between the runway and the security lights at the tops of these lights. During this whole procedure, I kept the throttles set at take-off thrust. Just before turning back for the fly-past, we flew through patchy light rain, which was of no significant concern, but as I dived back down to about 50 feet above the ground, the aircraft warning systems started to scream; frantically, "*Woop Woop* pull up!" "*Woop, Woop,* pull up!" "Gear, gear!" "Terrain, terrain!" "Too fast, too fast!" These warnings repeated as I continued in the dive.

As I flew past the terminal building, I looked out to see all the Sun Air staff standing in a line, in their familiar pink and blue uniforms, frantically waving us farewell. It was a touching scene, and as we passed abeam, I waggled the aircraft wings and pulled up in a high-nose attitude, which is unprecedented with passengers on board.

We now had the Johannesburg-based cabin staff on this returning leg. Compared to the atmosphere I had enjoyed with the Cape Town crew over the last two days, I found them very staid with more of a 'business as usual' matter of fact, demeanour.

After landing in Johannesburg, ATC directed us to a parking bay south of the terminal building. The passengers all bade us farewell and thanked us for an exciting flight. Then I saw the MD, Johan Borstlap and his wife standing at the foot of the stairs, looking very formal, and once all the passengers had disembarked, they came on board to thank the crew for their contribution.

However, compared to the previous night's celebration, this was a bit of an anti-climax.

So, I ended nearly twelve happy years with Sun Air, and once again, I was unemployed.

41 INTOLERANCE TO ALCOHOL

In other stories, I have related the problems I have experienced with debilitating hangovers, and even after reducing my drinking alcohol to two drinks, I still suffered excruciatingly. In more recent years, I learnt that even after a moderate amount of alcohol the hangovers were due, in no small measure, to an innate physiological problem.

In 1984, I discovered the reason for my problem by pure chance. I was then 42 years old and flying for Court Helicopters. By that stage of my life, I had stopped drinking alcohol and so was asked by a friend if I could participate as a control for a trainee doctor, who was doing a thesis on alcoholism. The original control for my age bracket was a pastor who had opted out before completing all the tests.

This series of tests included taking blood samples and ultimately undergoing a brain scan. As I also suffer from claustrophobia, I battled to keep myself from panicking while being pushed into the confines of the scanning machine. The only way I managed to keep the anxiety under control was to close my eyes and sing choruses to myself. I was also naïve about this new technology and thought the operator could somehow read my thoughts, which I endeavoured to keep pure! When it was all over, I learned that it had been the brain scan that had freaked out the pastor.

After I completed all the tests, our friend Mimi, also a doctor, told me that the results revealed that I was afflicted by what is known as Gilbert's Syndrome. Only five per cent of the world population has this condition, and most of these sufferers are women. She explained that it affected the ability of the haemoglobin in one's blood to retain blood sugar. This information immediately clarified why I struggled whenever I missed breakfast or went for a few hours without a substantial meal containing some form of protein. By 11 a.m., I would become weak, shaky, and even nauseous at times. Without knowing about this inbuilt disability, I had learnt to keep an emergency food supply, including ration biscuits (dog biscuits!), under my seat in the Alouette in case we were called out just before a meal. These callouts could sometimes run into hours if they developed into a contact. When I was the 'K-car' pilot, we often had to stay aloft for long periods to monitor and direct the troops deployed by the 'G-cars' and provide top cover.

Since retirement, I read on the Internet that Gilbert's Syndrome also causes an alcohol intolerance and makes one irritable when missing a meal. It also results in incapacitating hangovers, which in my case, would occur after only two drinks.

Joining the Air Force straight out of school, I assumed everyone experienced the same difficulty when introduced to alcohol without any previous exposure. I thought that if I persevered, I would be able to overcome this problem, especially with the macho Air Force saying, "If you can't drink, force yourself."

Once I received my commission and proudly wore the thin stripes indicating my rank as a junior Pilot Officer, I would need a good excuse to be absent from a Friday evening at the officer's mess pub. The Air Force had a tradition where they expected officers to show up at the Officers' Mess on Friday nights, referred to as 'pub night' or 'prayer meeting'. But I never felt at home in the noisy, smoke-filled atmosphere of the pub and would generally take up a bar stool near someone with whom I enjoyed talking. On most occasions, when I realised that I had over-indulged, I found it helped if I switched to Coca-Cola for a while before leaving.

Thornhill's working hours were from 0630 to 1330,

with every afternoon free to go home or play sports. These hours worked well, but sadly many of the young single officers developed the habit of going to the pub on Fridays, straight after work. Or those who did not play sports, or had no outside interest, would be in the pub every afternoon, without even bothering to have lunch. They would then return to their single-quarter rooms and fall into an intoxicated sleep before supper. Sadly, some of these same young officers became alcoholics and drank themselves to an early death.

In those days, when you walked into the pub, a thick pall of smoke filled the whole room and the adjoining snooker room. As I was not a smoker, this added to my discomfort, and the noise was almost deafening with raucous shouts and laughter, with officers playing darts, snooker, and liar-dice at the bar counter. One had to shout for anyone to hear one.

However, a few drinkers seemed to think it unmanly to sit on a barstool. One of these was Cyril White, who would stand all night quietly and not say much, nursing a pint-sized beer mug against his chest, while he surveyed the scene with an enigmatic smile of contentment.

Occasionally, I would forget about the next day's inevitable consequence when I was enjoying myself. My devil-may-come attitude would lead me to accept a cigarette from one of the continually proffered boxes, which would enhance my hangovers. Later, when I tried to close my eyes in bed, the room would start to spin, which made me feel extremely nauseous. When I did fall asleep, it would be a fitful night with a progressively worsening headache. The following day, I would suffer so badly that I could not sit up long enough to put on my socks and struggle to get to work on time.

It was during this time that Tish suffered an occasional blinding migraine headache. The Air Force doctor prescribed her effective little pink tablets to alleviate this. On the occasions that I could not make it out of bed, I took a quarter of her tablet, and in no time, all feeling would disappear from the top of my head to below my waist, which made it feel like a hollow vacuum. In this state, I experienced such relief from my symptoms that I found I could even do physical work in the sun without any problem.

I have already related many occasions when I have suffered hangovers that made it impossible to fulfil my duties. And one instance was when my flight engineer, Bob Mackie, who a few other pilots and I had unofficially taught to fly, managed to fly the helicopter to Mukumbura, completely unassisted, while I slept in the back.

Two similar experiences took place while I was on detachments to Kariba. On the first occasion, we had positioned to Kariba the day before security forces tasked us to fly SAS operatives to plant landmines across the Zambezi River, north of Chirundu. My flight engineer Brian Warren and I went to the Cutty Sark Hotel that night. Once again, I had far too much to drink as Brian had a good sense of humour, which always brought out the worst in me, as we tried to outdo each other with snappy and 'clever' repartee.

When we got back to the airfield that night, I knew I was in for it, as my head started splitting, and once again, the bed began its violent spinning. Despite taking a tablet before going to bed and trying to drink lots of water, I still woke up feeling like death warmed up.

Although I was supposed to lead the pair of helicopters to drop SAS troops, I was incapable of map reading to the correct spot. My only option was to ask John Annan, who was the epitome of what a professional pilot should be, if he would mind leading the formation.

I followed him like a robot without having enough strength to check if he was even taking us to the correct drop-off location.

Later that day, I believe that a bus travelling down the

road hit one of the landmines laid by the SAS members we had dropped. In my naivety, I was shocked when I heard Prime Minister, Ian Smith, denying on television that Rhodesian security forces had anything to do with this incident.

On the second occasion, I had stayed up all night drinking at the small pub at FAF 2 airfield and, at first light, I donned my overalls and walked straight out to the waiting helicopter with Sgt Phil Tubbs, my flight engineer.

Our assignment was to fly around to Lake Kariba's other side to swap a relay team stationed on one of the hills near Bumi Hills.

Knowing the condition, I was in and struggling to stay awake, I decided to take the shortcut. It would usually take about forty minutes to fly around the bulge to the east of the lake, as the Air Force banned us from overflying any stretch of water if we could not make dry ground in the event of an engine failure. On the other hand, flying directly to Bumi over a narrow section of the lake took only twenty minutes. After telling Phil my intention, I flew as high as I could to give us the best chance in the event of a mishap, and while I was flying, I viewed each island we passed as a potential emergency landing spot.

While I was making an approach to the LZ on the top of the hill, I felt so bad that I had to keep shaking my head to help me concentrate as I battled the overwhelming waves of tiredness. I desperately wanted to shut my eyes for a moment to give them a rest but knew just how dangerous it would be to allow myself this luxury.

In the end, I managed to complete the task, and returned to the hard-standing area at the FAF and headed straight off to bed.

Before I went off to sleep, I profusely apologised to Phil for putting his life at risk by flying in such a state of intoxication. I will never forget his reply, which was both humbling and complimentary when he said, "Sir, I would rather fly with you drunk, than a lot of the other pilots when they are sober!"

Years later, Al Bruce and I were captains in Sun Air and were assigned to ferry a DC-9 from Johannesburg to Shannon in Ireland. The trip took us one day, routing via Luanda, Bouaké in the Ivory Coast, and Marrakesh. After settling down in the hotel in Ireland, Al and I went through all the invoices we had incurred on the trip by spreading the American dollars and other paperwork across one of the beds in his room.

While we counted the money, he commented, "Twenty years ago, when we were young pilot officers in the RRAF, we would never have guessed that one day we would be sitting in a hotel room in Ireland, counting out thousands of US dollars!"

That evening some of our hosts took us to Bunratty Castle, where we were each greeted at the entrance by a gorgeous Irish girl in Medieval dress and a beautiful melodic Irish lilt with, "Good evening me Lords and Ladies,". The waiters handed us goblets of 'mead', which Al told me was 'pure plonk` when it was too late to make any difference!'

We then joined all the other guests at long banqueting tables in the shape of a T with the King and Queen at the head. It was a splendid evening, with lots of noise, while they continually topped up goblets with jugs of mead, which tasted like the nectar of the gods. While we ate joints of meat in our hands, enchanting Irish beauties serenaded us with haunting melodies and joyful tunes. One of our co-pilots, Craig Oosthuizen, who fell in 'love' with a red-headed beauty, was left behind at the end of the evening. I am not sure how he got back to the hotel, but I doubt he had much chance of striking up a relationship in such a strictly Roman Catholic country.

On the way back to the hotel, Al and I shared a taxi. I recall having some difficulty with my speech as I slurred out, "You know Al when you said that twenty years ago, we would never have guessed that one day we would be counting money in a hotel room in Ireland. Well, I've been working it out and think it was more like thirty years ago!!" Al, who has always managed to drink with little apparent effect, replied: "I was wondering how long it would take you to work that one out!" Luckily, I was too anaesthetised to take any offence.

The next day, Al and our co-pilots were due to leave early in the morning to catch a flight to Heathrow Airport and then a connecting flight back to South Africa. Meanwhile, I was booked onto a later flight to travel to JFK and connect to St Louis for our annual DC-9 refresher simulator training with Flight Safety International.

When I got to the hotel room, I realised that I only had to be at the airport just across from the car park by 9 o'clock the following day. So, I did not bother to ask for a wake-up call as I thought I would wake up with more than enough time to have a leisurely breakfast and wander across to the check-in counter.

Well, I woke up with another king-size headache and felt nauseous. When I looked at the time, I realised in total horror that I had fifteen minutes to get to the check-in desk. I still had to get dressed, pack, and check out of the hotel with no chance of a reviving shower. I hate to think what kind of an apparition I must have presented as I faced the fresh-faced Irish girl at the check-in desk that morning.

My wife's words still echo in my mind as I write: "You just never learn!" I suppose we all must contend with and learn to deal with different aspects of our physical makeup. But I am not sure if I am just a slow learner because it seemed to take a long time before I finally recognised and then eventually accepted my weaknesses and shortcomings.

43 TURBULENCE AND THUNDERSTORMS

Thunderstorm activity

Pilots have a healthy respect for thunderstorm activity and avoid it, especially when approaching land. A microburst has caused many accidents during thunderstorm activity near an airfield. This phenomenon is a concentrated column of air blasting down to the ground over a small area. Aircraft have sometimes flown into one of these and then battled against an uncontrollably high rate of descent.

Aeronautical technicians have advised that if pilots fly into microburst activity, they should immediately apply full thrust and pull a high nose-up attitude, just short of stalling the aircraft. However, safety officers have fed the same condition into simulators worldwide but, despite prewarning the pilots what to expect and what actions they should take, few of them have managed to avoid hitting the ground.

Although this situation is more often associated with the presence of roll-cloud activity ahead of an approaching cumulonimbus storm cloud, pilots treat every thunderstorm with a certain amount of trepidation.

Pilots use the weather radar to circumnavigate storm cells, which show up dark red on modern equipment. However, there are still times when a wall of thunderstorms lines up like a savage, barbaric horde, some distance on either side of the aircraft track, with just no way around them. Under these circumstances, cockpit crews make every effort to verbally warn the passengers and switch on the seat belt warning signs. They also discontinue all in-flight services. They then take a deep breath and steer the aircraft through the areas that show less dark red. Although one expects the storm centre to exhibit the most dramatic thunderstorms, this is not always the case. Often, an innocuous zone, in clear air directly beneath the characteristic 'anvil,' can be just as bad. Funnily enough, hail does not reflect on the radar screen, and pilots are often taken by surprise when the aircraft is unexpectedly pelted by drumming hailstones.

Certain areas worldwide are more prone to thunderstorm activity, and the tropics are particularly susceptible. Cloud tops sometimes extend well above sixty thousand feet AMSL in these latitudes. But unfortunately, airliners cannot climb above these heights, as designers have set most passenger aircraft to operate between thirty and forty thousand feet. Even outside the tropics, there are still specified areas that become known as 'thunderstorm alleys' where towering storm clouds start in a specific region and start their slow march, one after the other, out of the west. When these individual cells bunch up together, they give airline pilots the biggest headaches. Once the captain has no other options, he must just grit his teeth and push on straight through. He alone can decide and pray that the turbulence and other thunderstorm characteristics will not be too severe or overly protracted.

When captains are obliged to penetrate and weave through heavy thunderstorm activity areas, conditions can vary from moderate to severe turbulence, heavy rain, icing, hail, and lightning in varying combinations and degrees of intensity.

Sometimes ice can build up on the leading edges of the wings, engine intakes, and other probes, which sense the ambient air temperature and pressure very quickly. This ice can be so intense that it destroys both the wing's lift capability and the engine's power output and sometimes results in some alarming but spurious instrument indications. At the same time, the aircraft may be tossed around furiously, with blinding flashes of lightning and the petrifying and deafening assault of hailstones smashing against the windscreen and upper fuselage. Apart from dealing with the abovementioned problems, pilots must remain calm and collected.

On a few occasions, when I have entered what appears to be a run-of-the-mill, benign-looking cloud, bolts of lightning and claps of thunder, audible from inside the aircraft have assailed us. On one flight, after leaving Cape Town and setting the course for Johannesburg, we were still on the climb when a bolt of lightning struck us. We heard an ear-splitting clap around the DC-9. It felt like a giant conveyor belt had been slapped across the fuselage's top and caused the whole aircraft to bounce. As it hit, I thought it had struck the aircraft just above the cockpit, but the cabin staff were equally convinced it had struck above their section at the rear of the plane. Surprisingly, there was no visible or detrimental effect on the radio or electronic components of the aircraft. On other planes, I have seen scorch marks and small holes on the side of the fuselage, like the spattering from an arc welder.

At other times, turbulence can affect the aircraft under clear flight conditions, which occurred one night when our daughter, Kim, was a passenger on one of Bop Air's thirty-seat turboprop Brasilia aircraft. Clive Bradnick, a good friend of mine, was the captain from Johannesburg to Mmabatho.

Kim was coming to visit us with our six-month-old

grandson on her lap. Just as the air hostess was handing Kim a drink, the phenomenon slammed the aircraft so violently that passengers thought they had hit the ground. The ice cubes shot out of Kim's glass and landed on a passenger's lap behind her, and the air hostess let out a scream while our grandson shot out of Kim's arms and hit the overhead bulkhead. After a sympathetic apology from the captain, the flight continued unscathed.

On another night, I was flying with Rob Bester, a very pleasant co-pilot who had recently completed his training on the DC-9. We were on a flight from Cape Town to Durban, where Rob had lived all his life. The company had recruited and stationed him in his hometown, where the roster had scheduled me to spend the night at a beachfront hotel.

We had just commenced the descent with eighty nautical miles to go and could already see the lights of Durban as we passed the string of coastal towns to our right. After standard in-flight dinner service, the cabin crew was busy in the aisle clearing porcelain plates and glasses when the aircraft entered severe turbulence. It bounced and shook so violently that I could hardly read the instruments. The control column thrashed around as the aircraft continued to buck so frenetically that it automatically disengaged the autopilot. This out-of-control chaos lasted for endless minutes while we fought to reduce the speed and keep the wings level. The turbulence sent the crockery and glassware flying around the cabin in the back. At the rear, one young Indian cabin attendant, barely five feet tall, was catapulted upwards so violently that her head struck the upper bulkhead in between the overhead lockers. When she plummeted back to the cabin floor, she lay there in tears, like a rag doll. For the next few weeks, she could not work and had to wear a neck brace.

I had recently returned from simulator training in Zurich, where the flying pilot controlled the aircraft during the imposed emergencies. As Rob was more familiar with this localised weather phenomenon, it never entered my mind to take over the flight controls or show any outward sign of panic. I left him to handle the situation and suggested that he pull the nose up to reduce speed.

Once the aircraft had settled down, I reported the turbulence to the Durban ATC. Then, because we had raised the nose to keep the speed right down, we could not descend fast enough for a straight-in approach. To lose the extra height and configure the aircraft for a landing, I asked permission to carry out a right-hand orbit overhead the airfield and onto the base leg, giving the cabin staff more time to prepare the cabin for the landing.

Rob recognised this as a well-known phenomenon that forewarned a rapid approach of a southerly airflow and knew that the surface wind would soon swing around 180 degrees from its current northerly direction at any stage without warning. While we carried out the right-hand orbit onto the final approach, Rob suggested that the Tower Controller give us a constant readout of the surface wind. By the time we crossed the threshold, the wind had become becalmed, but the following SAA Airbus encountered such fierce turbulence just as it reached short finals, that it had to abandon the landing and do an overshoot procedure. The wind had swung completely around to come from the south as the turbulence hit. The reversal of wind direction forced the Airbus to reposition for the following approach from the north. Simultaneously, as we taxied into dispersal, the wind had whipped up into such a frenzy that the SAA aircraft had to abandon any further attempt to carry out a safe landing and had to divert back to Johannesburg. For the next few hours after that, ATC cancelled all flights in and out of Durban.

I was rather touched when Rob thanked me for trusting him enough to continue as the flying pilot throughout this tricky situation.

Later as I drove to the hotel, the gusting wind severely buffeted the car, and I noticed some trees bent almost double and others wholly uprooted.

The next day, Rob and I flew together again to leave from Durban to Johannesburg. As we entered Johannesburg airspace near the Vaal Dam, ATC instructed us to return to Durban, due to the thick fog that had rolled over the airfield, which had become so bad that no aircraft could get in or out.

Back on the ground in Durban, I called our local operations and asked them to phone the ops staff in Johannesburg to inform us the moment they could see the big grey transmitter south of the runway. This facility looks like a giant golf ball, which they could see from their office window. This 'ball' sat alongside the runway's final approach and housed one of the let-down facilities. I had previously noticed that you could see it from the operations room window. When it became visible, it was an indication that the visibility was improving and would soon exceed the 550 meters required to make a safe landing off a Cat 1 ILS approach.

As soon as I received the message that the ball was visible, I called Durban Tower for start-up clearance. In this way, I managed to get a jump start on all the other aircraft waiting to hear from the Met Department and Air Traffic network. By the time we reached Johannesburg, the weather had greatly improved, and no one knew on what grounds I had chosen to obtain air traffic clearance long before the official word had reached the other airline pilots.

Rob and I continued to handle bad weather conditions, had gained a lot of respect for each other's flying and operational ability, and went on to become good friends.

44 BOEING 727 FLYING AND NIGHT FREIGHT

Boeing 727

By December 1997, Sun Air had been in operation for nearly three years, rising out of Bop Air's ashes at the end of apartheid in South Africa in 1994. The company moved from its base in Mmabatho to Jan Smuts and evolved into a competitive commercial airline. Despite Comair's attempt to block the way, service soon started between Johannesburg and Cape Town, with a single DC-9 aircraft ZS-NNN, which we affectionately named 'Nancy.' Over the next two years, two more DC-9 aircraft joined the fleet. These were purchased from Iberian Airlines and refurbished by an Irish firm operating out of Shannon. Sun Air then expanded routes to include Durban and, for a while, Sun City and Livingstone.

The airline's reputation grew, and it became renowned for its friendly cabin attendants who served in-flight catering with good quality crockery and cutlery. Terry McKanick, the Cabin Services Manager, trained the cabin staff to meticulous grooming, professionalism, and charm.

The airline became extremely popular with

passengers, who enjoyed the high standard of service offered, and it won the award for the best national airline for three years in a row. The company decided to increase the frequency of scheduled flights between Johannesburg and Cape Town by leasing a Boeing 727 from Safair. As I was the longest-serving captain in the company at that time, I had the first option to convert to this beautiful three-engine aircraft.

The Boeing 727 first took to the skies in 1963 and came into service a year later. It was the most popular jetliner for a long time and proved phenomenally successful with airlines worldwide. It could land on shorter runways due to its unique wing design, where the flap extension and leading-edge slat deployment, almost doubled its wing surface area.

On 12 December 1996, I was one of two captains and a first officer, that Sun Air sent to undergo lectures and simulator training with Flight Safety International in Pittsburg, Pennsylvania, in the USA. The other captain was Arthur Downes, who had joined Bop Air direct from the Anglo-American aircraft division in Johannesburg when we had purchased the Gulfstream III. He was already in his mid-fifties and had spent several years flying this same Gulfstream before Anglo upgraded to the Gulfstream IV. He was a gentleman, who loved good food and was always half-heartedly trying to lose weight with different diets. He had a certain charming humility, which made him popular among the young co-pilots, and he was ready and willing to fly whenever asked as he felt that any prolonging of his flying career was a bonus. When Sun Air introduced the first DC-9, he had won enough respect to be one of the first to go for training.

The first officer was Glen Watson, who had come straight off Mirage IIIs in the SAAF. He was still single in his mid-thirties but charmed all the cabin staff with his sense of humour.

Safair also sent along a couple of their flight engineers

with us. They upgraded from the C130 Hercules four-engine turboprop aircraft to the Boeing 727. These engineers related amazing stories of trips they had carried out on the C130 aircraft to Sudan and East Timor for the UN's World Food Program. They had also carried out some scary flights from southern Argentina to the Antarctic to resupply the South African base. That region's unpredictable and sudden weather changes made it tricky to cross the Southern Ocean with sufficient fuel to carry out a diversion if necessary. On occasions, they would have to land on a snow-cleared runway in the most precarious conditions while experiencing a 'white-out' in a blizzard.

On our trip from Johannesburg to Pittsburg, we flew in business class on South African Airways. But, although the seats had more room than in the economy class cabin, they did not recline enough to suit my needs, and I still found it difficult to sleep while I was half sitting up. In the end, my back was so uncomfortable that I knelt on the floor and slept draped across the seat. The following day several people who had witnessed this unusual posture commented that they thought I was praying all night!

Our hotel was not far from Pittsburg airport, and there was a well-stocked buffet restaurant next door. It was the first time I had ever seen such a collection of obese individuals, who could hardly walk properly but still managed to pile excessive amounts of food onto their plates. Arthur enjoyed this facility, and I would often join him at mealtimes.

Flight Safety undertook our training in their usual professional manner, and I quickly took to the aircraft's handling characteristics, with its big 'man-sized' control wheel. There was a small amount of play on either side of the control wheel's neutral position in the rolling plane, where no immediate response occurred. However, once accustomed to it, the aircraft had a rock-solid and stable feel. I enjoyed having three engines and the interaction between the pilots and the flight engineer, who sat behind the pilots, facing starboard at his console. He had his area of responsibility for the fuel and other engine operating systems. Sadly, aircraft have advanced to such an extent that computers have taken over this function. I also enjoyed Chuck, our flying simulator instructor, who would sit in the jump seat directly behind me in the cockpit, with a very comforting way of relating to me by occasionally placing a hand on my shoulder and referring to me as "Big Fella."

Our last day of training was just a few days before Christmas. The hotel ferry bus was a bit late in coming to collect us. So, as it had snowed most of that day, while we waited, we must have looked like a bunch of kids as we started to run around in the crisp white snow, pelting each other with snowballs.

We arrived back in South Africa, just in time to celebrate Christmas with our families. On 28 December, I went for a 727-flight check with Cyril Rogers, a previous SAA training captain. After retirement, he had a short spell with Air Mauritius before joining Bop Air. He was a slight, dignified gentleman with a neat, white moustache and a clipped, almost British-speaking manner. He could have slotted right in among the Battle of Britain pilots of WWII. He had flown the Boeing 727 when SAA had first introduced them to their fleet some years before.

After a bit of handling practice away from Johannesburg flight traffic, we joined the circuit for some landings and takeoffs on runway Zero Three Right. Initially, I had trouble with the landing technique, so I asked Cyril to demonstrate a circuit. I had often done this in my flying career, which had always proved extremely helpful, as most instructors enjoy the opportunity to do some 'hands-on' flying themselves.

As he neared touchdown, I noticed an unexpected

method, which I am certain Cyril would not have been able to explain since instructors often become so familiar with an aircraft that their control inputs become conditioned reflex actions. I noticed that he held a constant nose attitude right down to the runway. Then, instead of flaring to check the descent rate, he would push forward on the control wheel to lower the nose into an almost level attitude until the aircraft slid effortlessly onto the runway. I employed this same technique on my next landing and found it worked like a charm. After that, I never experienced a hard landing on the 727, which boggled many flight engineers, testing officers, and co-pilots with whom I flew later.

One night, we arrived in a drizzle on a flight from Cape Town to Johannesburg. The runway was wet, and I used Cyril Roger's method to land. Nobody was aware that the aircraft was on the ground until I applied reverse thrust, and the plane started to decelerate. When the passengers disembarked, every single one of them popped their heads into the cockpit to compliment me on the smoothest landing they had ever experienced.

In contrast, I had battled to carry out smooth landings on the DC-9, as its narrow wings have no cushioning effect whatsoever, so being able to master the 727 landings did much to restore my self-esteem. I later analyzed this unfamiliar landing method and concluded that it was due to the large wing area of the 727, which allowed it to build up a ground cushion as it got close to the runway. When the pilot pushed the nose forward, it further compressed this trapped air, making it possible to make a smooth landing every time. Later, when I flew the 727-100 model with a shorter fuselage, I found it required the more usual technique.

After three months of flying the 727 between Johannesburg and Cape Town, Safair acquired four MD-80s which they leased to Sun Air, who then discontinued leasing the 727s. A fifth MD-80 followed about a year later, and

Sun Air took the three pilots off the 727s. After a quick re-familiarisation with the DC-9 aircraft, we returned to scheduled flights, while the company sent the other pilots to Zurich for simulator training on the MD-80 aircraft. I later followed with our girl pilot, Maxie, and found it a fascinating city. While there, it amused me to see locals in a pub enjoying Mr Bean on TV. His mimed humour managed to break all language and cultural barriers.

Maxie and I did some sightseeing, and at one café, a black man served us. When I asked him where he came from, he said South Africa. So, I asked him which was his tribe. He replied, "Zulu." As I know a smattering of Zulu, I thought I would impress him and make him homesick, but when I spoke to him in Zulu, he just stared at me blankly. So, I asked him where in Natal he was born. Once again, he just stared at me in total confusion, which made me realise that he must have gained considerable sympathy over the years by playing a downtrodden victim of apartheid.

Then, in 1999, came the shock when SAA bought and liquidated Sun Air the next day. As I was already fifty-seven, SAA and British Airways had no interest in employing me, but thankfully, with my 727 rating, I was immediately offered employment by Safair.

After another quick refresher flight with Robbie Robinson, a previous DCA flight examiner who flew for Safair. Once again, I impressed everyone with my smooth landing technique.

After my check ride, Safair put me online to do night freight to Cape Town and East London, with stop-offs in Durban and Port Elizabeth. During the first six months, I was fortunate enough to do daytime flights to Zimbabwe, Zambia, and Malawi on a contract with flower growers, who exported their produce to Europe from Harare, using a chartered DC-10 aircraft. I also had regular flights on the 727 for Comair's scheduled flights to Cape Town and a few charter flights on the Safair MD-80 aircraft. Apart from these occasional flights, these aircraft sat sadly in the parking area for months on end.

A welcome break came when I was sent with Captain Mike Kemp to Vienna to collect an MD-82 for Safair. He was another highly qualified training captain from SAA, who had joined Bop Air and continued as the chief training captain until it became Sun Air. We spent a week in Austria while a technical company refurbished and painted the aircraft for export. We then flew it back to South Africa, with a refuelling stop in Cairo and a night stop in Nairobi, and Mike used this flight to carry out my annual instrument rating check.

My last year in Safair fell into a monotonous routine of flying night freight, three or four nights a week and then having the rest of the week off. During this time, I tried to remain positive but struggled with the continual alteration to my sleep pattern. Whereas many pilots who do night freight seem to use alcohol to help them sleep on arrival at the hotel in the early hours of the morning, I thought that was not a particularly good idea. It had almost become a tradition for some of the 'old school' captains to volunteer the co-pilot's hotel room as the place to meet. Despite having checked into the crew hotel after 4 a.m., most flight crews would extend the meeting until morning to make sure they could indulge in breakfast in the dining room, as the company paid for it.

When I started, I made it known that I intended to go straight to bed from the outset. Once I had showered, I would have some biltong, cheese, rooibos tea, and possibly yoghurt and fruit before falling into bed with an eye mask. I would then place a 'Do Not Disturb' sign on the door handle, and after taking a sleeping tablet, I would set the alarm for midday. When I woke, I would then have a quick shower and some more of my rations before jogging to the gym for a workout. I would end off by spending time in the sauna before another shower and then return to my room for an afternoon sleep in preparation for that night's ordeal.

Invariably, two Safair 727 freight aircraft would arrive in Johannesburg from Cape Town and East London at around 2300 hours. The crews would all congregate in Safair's crew room or operations room to wait for a couple of hours while the handlers were loading freight onto the aircraft for the next departure. During this time, aircrew would moan about life over endless cups of coffee and cigarettes. Being new on the block, I had extraordinarily little in common with the other captains, most of whom were well into their sixties and needed to keep flying although hating their job.

I found the whole scene depressing, mainly because it was unnatural to stay awake when my inner time clock was screaming for me to go to bed. The worst part about it was that I never gave my body a chance to get into a settled sleeping pattern, which I felt was not particularly good for one's mental or physical health. The only way I could cope was to take sleeping tablets, which had some side effects, although the ones our local pharmacist had subscribed were relatively mild.

To keep myself occupied during the waiting hours, I asked Tracey, who was one of the operation's staff, to give me the password for her computer to relax and play mindless card games on it. She was an attractive girl, with long auburn hair reaching below her waist, and trusted me not to abuse the privilege.

The atmosphere and negativity would reach their lowest point when both aircraft's crew clambered into the same crew bus to the flight line opposite the freight hangar. I tried my best to avoid being pulled down by this atmosphere and remember one captain moaning about the unhelpful treatment he had recently encountered at his bank. He was a particularly sour-faced individual, and when I said that I found it helped to smile when dealing with bank ladies, I might just as well have been speaking a foreign language by the blank look of bemusement on his face. I elaborated that when boarding passengers blocked me from entering the cockpit, I noticed that although the cabin attendants welcomed everybody on board with a warm smile, most lady passengers responded with a smile. Few men did. Unfortunately, he was in no mood to consider this observation and continued his grievances.

After being with Safair for twenty months, towards the end of 2000, I decided I had enough of freight flying at night and had earned enough money to retire from flying. So, at the age of fifty-eight, and much to Tish's horror, I resigned from Safair and ended my forty-year flying career.

Printed in Great Britain
by Amazon